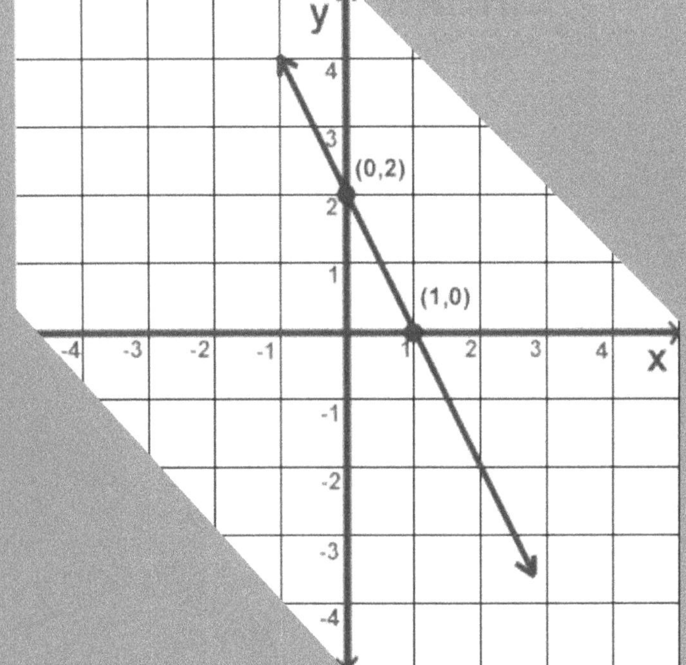

LINEAR ALGEBRA

MARY JANE G. PACAÑOT
IRENE C. MUTIA
MANILYN PERICANO
ELERY B. LEYBA
CHRISTIAN CABEN M. LARISMA

LINEAR ALGEBRA

Mary Jane G. Paca Ñot
Manilyn Pericano
Christian Caben M. Larisma
Irene C. Mutia
Elery B. Leyba

ISBN 978-621-8307-00-1

Published by Yawman Book Publishing House

Davao City Philippines

+639219512458; yawmanresearch@gmail.com

Acknowledgements

Linear Algebra would not have been made possible without the following, for which it is dedicated:

To our loved one's the source of our inspiration in this endeavor.

To Filipino students and Filipino teachers to whom this book is intended.

We say thank you to all authors and writers of different materials, references, and sourcebooks whose ideas and work has become an important part of this book.

We appreciate fellow writers who have greatly contributed to this labor of love, realizing and fulfilling what was once a dream.

We thank our professors for the invaluable support and inspiration for this work to push through.

To GOD, who graciously sustained our efforts and allowed us to become better teachers through this endeavor, be all Glory.

Contents

INTRODUCTION

Linear algebra is a vital course for students. Few subjects can claim to have such widespread applications in other areas of mathematics-multi variable calculus, differential equations, probability, and in physics, biology, chemistry, economics, finance, psychology, sociology. and all fields of engineering. It also presents the students with an excellent opportunity to learn and how to deal with abstract concepts.

This module introduces the basic ideas and computational techniques of linear algebra. It also includes a wide variety of carefully selected applications. It also introduces the students to working with abstract concepts. Finally, in covering the basic ideas of linear algebra, the abstract ideas are carefully balanced by a considerable emphasis on the geometrical and computational aspects of the subject.

Throughout this module, a great variety of examples and exercises of the important concepts are included. The study of such examples is of fundamental importance. It tends to minimize the number of students who can repeat definition, theorem, and proof logically without grasping the meaning of the abstract concepts. Assessments are also includedin every section of this module to measure the students' learning as to how far they have learned the concepts. Finally, answers on exercises are provided for the students to answer the assessments.[1]

As with the rest of the module, the exercises and assessments aim to build ability and help students experience the pleasure of doing mathematics. Studentsshould see how the ideas arise and should be able to picture themselves doing the same type of work.[2]

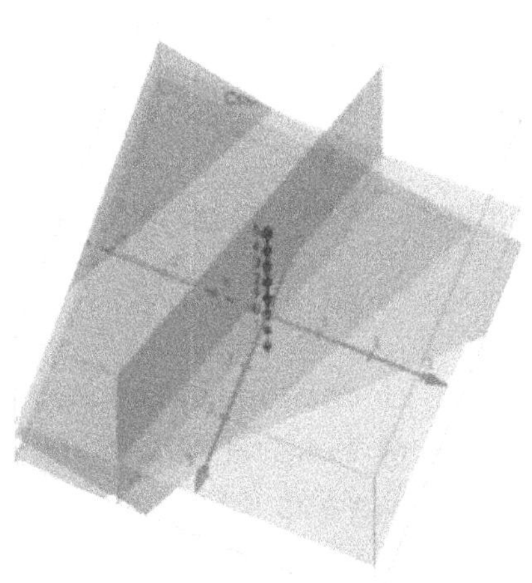

SYSTEM OF LINEAR EQUATIONS & MATRICES

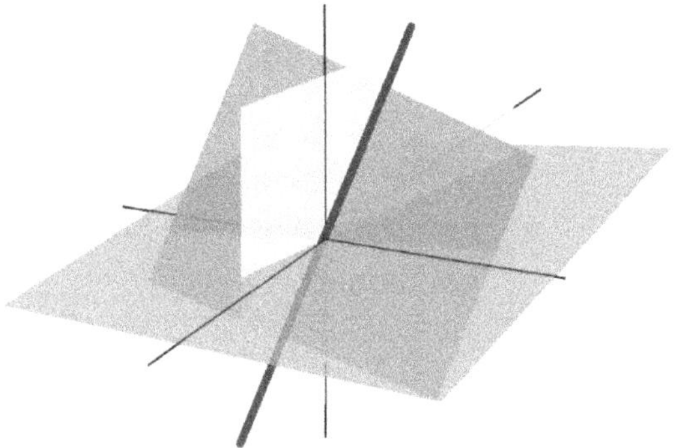

source: https://slideshare.net/hardisukhadia3/systems-of-linear-equation

LINEAR EQUATIONS AND MATRICES

OBJECTIVES

After working through this section, you should be able to:

- Understand the connection between the solutions of systems of simultaneous linear equations in two and three unknowns and the intersection of lines and planes in R 2 and R 3
- Describe the three types of elementary operation;
- Use the method of Gauss–Jordan elimination to find the solutions of systems of simultaneous linear equations;
- Perform the matrix operations of addition, multiplication, and transposition.
- Recognize the following types of matrix: square, zero, diagonal, lower-triangular, upper-triangular, identity, symmetric.
- Express a system of simultaneous linear equations in matrix form.
- Learn the algebraic properties of matrix operations and partitioned

1.1 SYSTEM OF LINEAR EQUATIONS

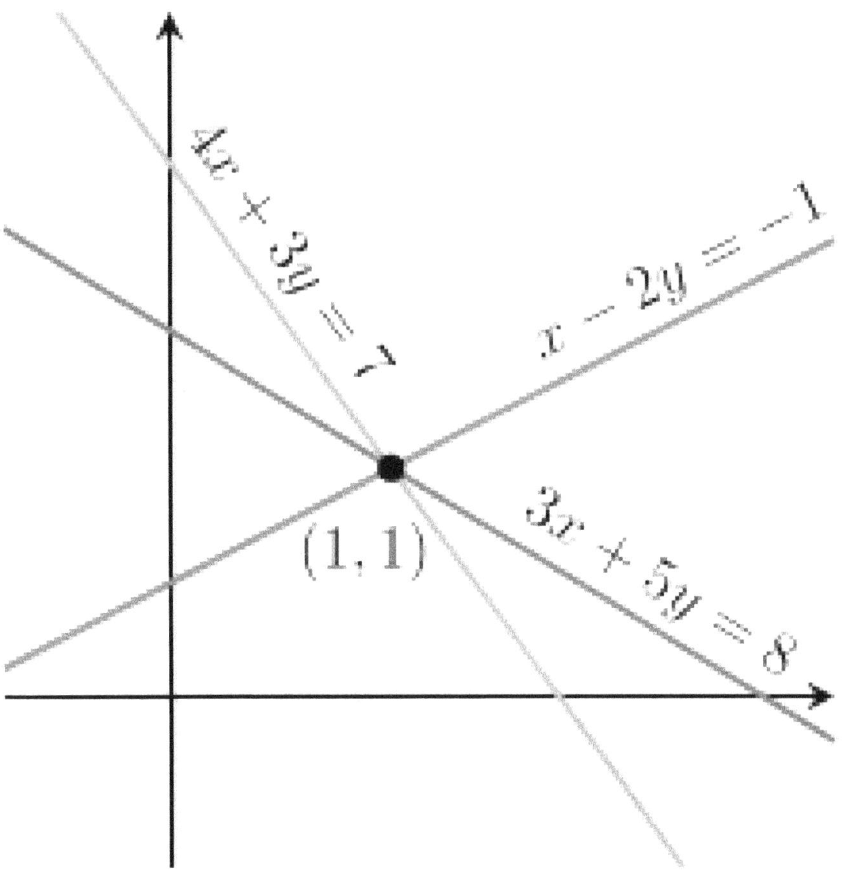

INTRODUCTION

Systems of linear equations can be used to solve resource allocation problemsin business and economics. The introduction to this chapter illustrates that there is more than one way to think of the solution to a system of linear equations. Thissection focuses on the following: Mathematical understanding; Insights from studies on linear algebra, Methods for solving a system of linear equations using matrix techniques. [1]

Systems of linear equations play an important and motivating role in the subject of linear algebra. Many problems in linear algebra reduce to finding the solution of a system of linear equations. Thus, the techniques introduced in this chapter will apply to abstract ideas introducedlater. On the other hand, some of the abstract results will give us new insights into thestructure and properties of systems of linear equations.[2]

DISCUSSION

A system of linear equations is a finite set of linear equations, each with the same variables.

A linear equation in the variables $x_1 \ldots\ldots x_n$ is an equation that can be written in the form:[3]

Where b and the coefficients $a_1, \ldots a_n$ are real or complex numbers, usually known inadvance. The subscript ***n*** may be any positive integer. In textbook examples and exercises, ***n*** is normally between 2 and 5. In real-life problems, ***n*** might be 50 or 5000,or even larger.[4]

The equations $4x_1 - 5x_2 + 2 = x_1$ and $x_2 = 2(\sqrt{6} - x_1) + \overline{x_3}$ are both linear because they can be rearranged algebraically as in equation (1).

$$3x_1 - 5x_2 = -2 \text{ and } 2x_1 + x_2 + x_3 = 2\sqrt{6}.$$

The equations $4x_1 - 5x_2 = x_1x_2$ *and* $x_2 = 2\sqrt{x_1} - 6$ are not linear because of thepresence of x_1x_2 in the first equation and $\sqrt{x_1}$ in the second. A system of linear equations (or a linear system) is a collection of one or more linear equations involvingthe same variables—say, $x_1 \ldots\ldots x_n$.[4]

An example is $4x_1 - 5x_2 + 2x_3 = 8$ $\qquad$ $x_2 - 5x_3 = 3$

A solution of the system is a list $s_1, s_2 \ldots\ldots. s_n$ of numbers that makes each equation atrue statement when the values $s_1, \ldots \ldots . s_n$ are substituted for $x_1 \ldots \ldots x_n$ respectively. The **set of all possible solutions** is called the **solution set of the linear system**.Two linear systems are called equivalent if they have the same solution set. Each solution of the first system is a solution of the second system, and each solutionof the second system is a solution of the first. Finding the solution set of a system oftwo linear equations in two variables is easy because it amounts to finding the intersection of two lines. A typical problem is.[3,5]

$$x_1 - 2x_2 = -1$$

$$-x_1 + 3x_2 = 3$$

The graphs of these equations are lines, which we denote by l_1 and l_2 . A pair of numbers (x_1 , x_2) satisfies *both* equations in the system if and only if the point. (x_1 , x_2)lies on both l_1 and l_2. As you can easily verify, the solution is the single point(3 ,2) in the system above. See Figure 1.[4]

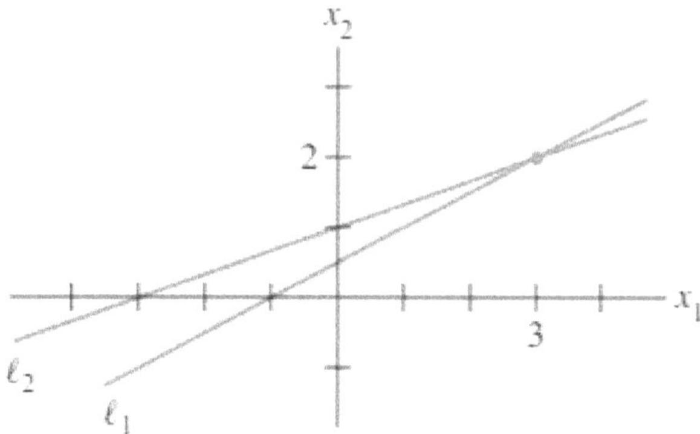

FIGURE 1 Exactly one solution.

Of course, two lines need not intersect in a single point—they could be parallel or coincide and hence *"intersect"* at every point on the line. Figure 2 shows the graphs that correspond to the following systems.[4]

$$x_1 - 2x_2 = -1, \qquad - x_1 + 2x_2 = 1$$

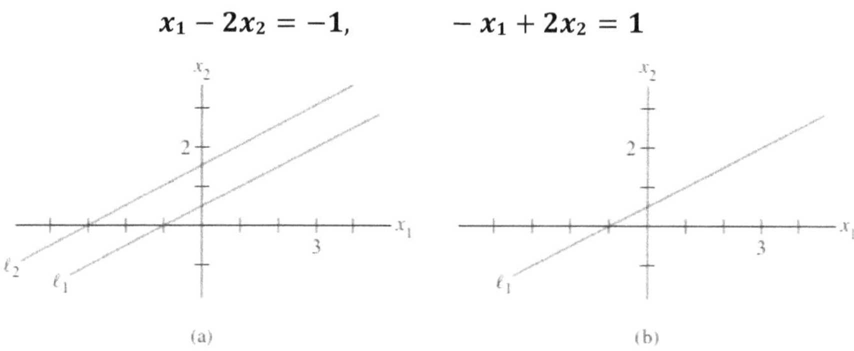

FIGURE 2 (a) No solution. (b) Infinitely many solutions.

A system of linear equations has

- no solution, or
- infinitely many solutions

A system of linear equations is **consistent** if it has either one solution orinfinitely many solutions; a system is **inconsistent** if it has no solution.[6]

SOLVING SYSTEM OF LINEAR EQUATIONS

We know that solving equations with one unknown—like $2x + 4 = 7x$, for instance—requires manipulating both sides of the equation until the unknown variable is isolatedon one side. For this instance, we can subtract $2x$ from both sides of the equation to obtain $4 = 5x$, which simplifies to $x = 4/5$.

What about the case when given two equations and must solve for two unknowns?

For example,

$$x + 2y = 5$$

$$3x + 9y = 21$$

Can you find values of x and y that satisfy both equations?

Concepts

- x, y: the two unknowns in the equations
- eq1, eq2: a system of two equations that must be solved simultaneously.These equations will look like

$$a_1x + b_1y = c_1$$

$$a_2x + b_2y = c_2$$

where $\boldsymbol{as\,, bs\ and\ cs\ are\ given\ constants}$

Principles

If you have n equations and n unknowns, you can solve the equations simultaneouslyand find the values of the unknowns. There are several different approaches for solving equations simultaneously. We will learn about three different approaches in this section.

SOLUTION TECHNIQUES

When solving two unknowns in two equations, the best approach is to eliminate one of the variables from the equations. By combining the two equations appropriately,we can simplify the problem to find one unknown in one equation.

1. SOLVING BY SUBSTITUTION

We want to solve the following system of equations:

$$x + 2y = 5$$

$$3x + 9y = 21$$

We can isolate x in the first equation to obtain.

$$x = 5 - 2y$$

$$3x + 9y = 21.$$

Now substitute the expression for x from the top equation into the bottom equation:

$$3(5 - 2y) + 9y = 21.$$

We just eliminated one of the unknowns by substitution. Continuing, we expand the bracket to find $15 - 6y + 9y = 21,$ **or** $3y = 6$. We find $y = 2$, but what is x? Easy. To solve for x, plug the value $y = 2$ into any equations we started from. Using the equation$x = 5 - 2y$, we find $x = 5 - 2(2) = 1$.

2. SOLVING BY SUBTRACTION

Let us return to our set of equations to see another approach for solving:

$$x + 2y = 5$$

$$3x + 9y = 21$$

Observe that any equation will remain true if we multiply the whole equation by someconstant. For example, we can multiply the first equation by 3 to obtain an equivalentset of equations: $3x + 6y$

$= 15$

$$3x + 9y = 21$$

Why did I pick three as the multiplier? The x terms in both equations now have the same coefficient by choosing this constant. Subtracting two true equations yields another true equation. Let's subtract the top equation from the bottom one:

$$3x - 3x + 9y - 6y = 21 - 15$$

$$or\ 3y = 6.$$

The $3x$ terms cancel. This subtraction eliminates the variable x because we multiplied the first equation by 3. We find$y = 2$. To find x, substitute $y = 2$ into oneof the original equations:

$$x + 2(2) = 5,$$

from which we deduce that
$x = 1$.

3. SOLVING BY EQUATING

There is a third way to solve the equations:

$$x + 2y = 5$$

$$3x + 9y = 21$$

We can isolate x in both equations by moving all other variables and constants to theright-hand sides of the equations: $x = 5 - 2y$,

$$X = \frac{1}{3}(21 - 9y) = 7 - 3y$$

Though the variable x is unknown to us, we know two facts about it: x is equal to $5 - 2y$ and x is equal to $7 - 3y$. Therefore, we can eliminate x by equating theright-hand sides of the equations:

$$5 - 2y = 7 - 3y.$$

We solve for y by adding $3y$ to both sides and subtracting five from both sides. We find $y = 2$ then plug this value into the equation $x = 5 - 2y$ to find x. The solutions are $\boldsymbol{x = 1 \text{ } and \text{ } y = 2}$.

The three elimination techniques presented here will allow solving any system of $\boldsymbol{n}$ linear equations in $\boldsymbol{n}$ unknowns. Each time you perform a substitution, a subtraction, or an elimination by equating, you simplify the problem to a problem of finding $(\boldsymbol{n} - \mathbf{1})$ unknowns in a system of $(\boldsymbol{n} - \mathbf{1})$ equations.

SOLVING SYSTEM OF LINEAR EQUATIONS IN TWO VARIABLES

A linear equation in one variable is an equation that can be written in the form $ax + b = 0$ where *a* and *b* are real numbers with $a \neq 0$. This definition can be extendedto more variables as follows:

Definition: A linear equation in *n* variables is an equation that can be written in theform$a_1x_1 + a_2x_2 + \cdots + a_nx_n = b$ for variables $x_1, x_2, \ldots x_n$ and real numbers

$a_1, a_2, \ldots a_n, b$ where at least one of $a_1, a_2, \ldots a_n$ is nonzero.

For example, $2x - 5y = -7$ is a linear equation in two variables while $4x_1 + 9x_2 - 17x_3 = 11$is a linear equation in three variables. The important thing to remember isthat all variables of a linear equation have an exponent of 1.[7]

Definition: A system of linear equations in two variables is the collection of two linearequations considered simultaneously. The solution to a system of equations in two variables is the set of all ordered pairs for which *both* equations are true.[7]

Consider the following three systems of linear equations in two variables.

$$3x - 2y = -9$$

$$x + y = 2 \qquad -3x_1 + 2x_2 = 11$$

$$8x_1 - 9x_2 = -2 \qquad \sqrt{2}a + b = \pi$$

$$-4a + \sqrt{3}b = 1$$

Notice in each of the three systems above that different variables were used. It does not matter what we call the two variables, provided that both equations are linear. According to the definition of a system of linear equations in two variables, the solution is the set of all ordered pairs for which both equations are true.7

If a linear system has at least one solution, it is consistent, and the solution is the set of all ordered pairs that satisfy both equations. If the system does not have a solution, it is inconsistent.

The solution to a system of two linear equations involving two variables can be viewed geometrically. Since the graph of each system equation is a line, we can sketch the two lines and geometrically view the solution. There are only three possibilities for the graph of a linear system of equations involving two variables:

1. The lines can have different slopes and thus intersect at a single point (Figure 1a).The solution is the ordered pair representing the point of intersection. This isconsistent since it has a solution. The equations are said to be independent.
2. The two lines can be parallel (have the same slope but different *y*-intercepts) and thus have no intersecting point (Figure 1b). This system is said to be inconsistent and has no solution. The equations are independent.
3. The two equations can be different representations of the same line and have the same graph (Figure 1c). In this case, there are infinitely many solutions. Because thissystem has at least one solution, it is consistent. The equations are said to be dependent.

Figure 1

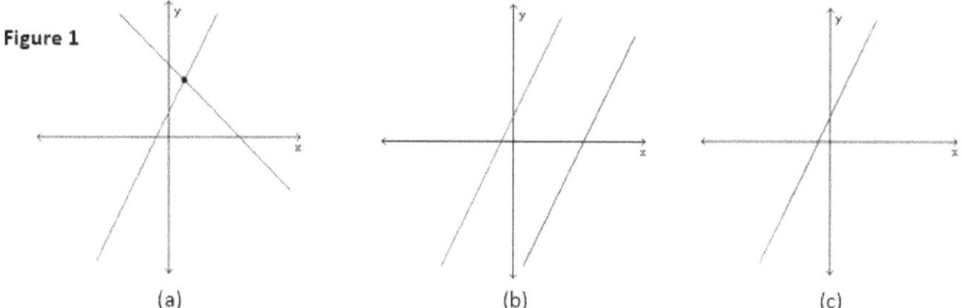

We will learn two algebraic methods used for solving linear systems involving two equations and two variables. These methods are known as the ***substitution method*** and the ***elimination method.***

1. Substitution Method

Solving a System of Equations by the Method of Substitution

Step 1: Choose an equation and solve for one variable in terms of the other variable.

Step 2: Substitute the expression from step 1 into the other equation.

Step 3: Solve the equation in one variable.

Step 4: Substitute the value found in step 3 into one of the original equations to find the value of the other variable.

Example: $x - 2y = 4$

$3y + x = 19$

Step 1: Solve the first equation for x.

$x - 2y = 4$

$+2y \quad + 2y$

$\boldsymbol{x = 4 + 2y}$

Step 2: Plug in $4 + 2y$ for x in the equation that we did not use.

$3y + x = 19$
$3y + (4 + 2y) = 19$
$5y + 4 = 19$

Step 3: Solve for y

$5y + 4 = 19$

$-4 \quad -4$
$5y/5 = 15/5$
$\boldsymbol{y} = \mathbf{3}$

Step 4: Go back to one of the original equations and solve for x using the y-value wejust found. For this example, we will use the first equation to solve for x.[8]

$x - 2y = 4$
$x - 2(3) = 4$
$x - 6 = 4$
$+6 \quad +6$
$\boldsymbol{x} = \mathbf{10}$

*Therefore, the solution to this system is **x = 10** and **y = 3.***

2. Elimination Method

Another method used to solve a system of two equations is the elimination method. The elimination method involves adding the two equations together to eliminate one of the variables. To accomplish this task, the coefficients ofone of the variables must differ only in sign. This can be done by multiplying one or both equations by a suitable constant. The elimination method can be summarized in the following five steps.

Steps for solving system of equations by elimination

Step 1: Choose a variable to eliminate.
Step 2: Multiply one or both equations by an appropriate nonzero constant sothat the sum of the coefficients of one of the variables is zero.

Step 3: Add the two equations together to obtain an equation in one variable.

Step 4: Solve the equation in one variable

Step 5: Substitute the value obtained in step 4 into one of the original equations to solve for the other variable.

Example:

$$2x + 3y = 14$$
$$3x - 5y = 2$$

Step 1: Even though we could pick any variable to eliminate, let's eliminate x in thisexample.

Step 2: To eliminate x, we will multiply through the first equation by **3** and the second equation by **2** to get the coefficients to be the same.

$$(2x + 3y) \cdot 3 = (14) \cdot 3$$
$$(3x - 5y) \cdot 2 = (2) \cdot 2$$

$$6x + 9y = 42$$
$$6x - 10y = 4$$

Step 3: Subtract the equations.

$$6x + 9y = 42$$
$$-(6x - 10y = 4)$$
$$0 + 19y = 38$$

Step 4: Solve for y

$$\frac{19y}{19} = \frac{38}{19} = 2$$

Step 5: Go back to one of the original equations and solve for x using the y-valuewe just found. For this example, we will use the second equation to solve for x.

$3x - 5y = 2$

$3x - 5(2) = 2$

$3x - 10 = 2$

$+10 + 10$

$3x = 12$
$\mathbf{x = 4}$

Therefore, the solution to this system is $x = 4$ and $y = 2$.

SYSTEMS OF LINEAR EQUATIONS IN THREE VARIABLES

Steps:

1. Using two of the three given equations, eliminate one of the variables.

2. Using a different set of two equations from the given three, eliminate the same variable you eliminated in step one.

3. Use these two equations (which are now in two variables) and solve the system.

4. Use the values you find in step 3 to find the third variable using one of the original equations.[8]

Example:

Solve the following system:
$$x + 2y + z = 10$$
$$2x - 6y - z = -15$$
$$3x + 2y + 2z = 17$$

Step 1: Use the first and second equation to eliminate the variable z.

$$x + 2y + z = 10$$
$$+(2x - 6y - z = -15)$$
$$\mathbf{3x - 4y = -5}$$

Step 2: Use the second and third equation and eliminate the variable z.

$$(2x - 6y - z) \cdot 2 = (-15) \cdot 2$$
$$3x + 2y + 2z = 17$$

$$4x - 12y - 2z = -30$$
$$+(3x + 2y + 2z = 17)$$
$$7x + 10y = -13$$

Step 3: Use the two equations we just computed to solve for x and y.

$$3x - 4y = -5$$
$$7x - 10y = -13$$

- We could use either elimination or substitution. For this example, let's useelimination to eliminate $\boldsymbol{x}$.

$(3x - 4y) \cdot 7 = (-5) \cdot 7$
$(7x - 10y) \cdot 3 = (-13) \cdot 3$
$21x - 28y = -35$
$-(21x - 30y = -39)$
$2y = 4$
$\boldsymbol{y = 2}$

- We will find x by plugging $y = 2$ into $\boldsymbol{3x - 4y = -5}$.

$3x - 4y = -5$

$3x - 4(2) = -5$
$3x - 8 = -5$
$+8 \quad + 8$
$3x = 3$
$\boldsymbol{x = 1}$

Step 4: Using $x = 1$ and $y = 2$, we will use the first equation in the original system ofthree equations to solve for z.

$x + 2y + z = 10$
$1 + 2(2) + z = 10$
$5 + z = 10$
$-5 \qquad - 5$
$\boldsymbol{z = 5}$

EXERCISE 1.1

1. Solve the system of equations simultaneously for x *and* y: $2x + 4y = 16$, $5x - y = 7$.

2. Solve the system of equations for the unknowns x, y, *and* z: $2x + y - 4z = 28$, $x + y + z = 8$, $2x - y - 6z = 22$.

3. Solve for p and q given the equations $p + q = 10$ *and* $p - q = 4$.

4. Solve the following system of linear equations: $2x + 4y = 8$, $4x + 3y = 11$.

5. Solve the system by substitution: $\begin{cases} 2x+y=7 \\ x-2y=6 \end{cases}$

6. Solve the system by elimination: $\begin{cases} 3x+y=5 \\ 2x-3y=7 \end{cases}$

7. What is the solution (x, y) to the given system of equations.

$$3x + 2y = 10$$

$$2x - 2y = 5$$

7. Is (8, −2) the solution for (x, y) to the system of equations.

$$5x + 3y = 34$$

$$3x + 3y = 30$$

8. Solve the given system of equations by the addition method.

$$3x + 5y = -11$$

$$x-2y=11$$

9. Solve the given system by elimination:

$$4x + 3y = 22$$

$$-5x - 4y = -28$$

1. Solve:

$$4x + 3y = -2$$
$$8x - 2y = 12$$

2. Solve the given system by elimination:

$$4x + y = 23$$

$$3x - y = 12$$

3. Solve the following system of equations using the substitution method:

$$x - \ -5y = 36$$
$$2x + y = \ -16$$

a. $14x + 7y = 217$
$14 \ + 3y \ = 189$

b. $x - 2y = 3$
$4x - 8y = 12$

c. $3x + 2y = 9$
$5x - 3y = 4$

1.2 MATRICES

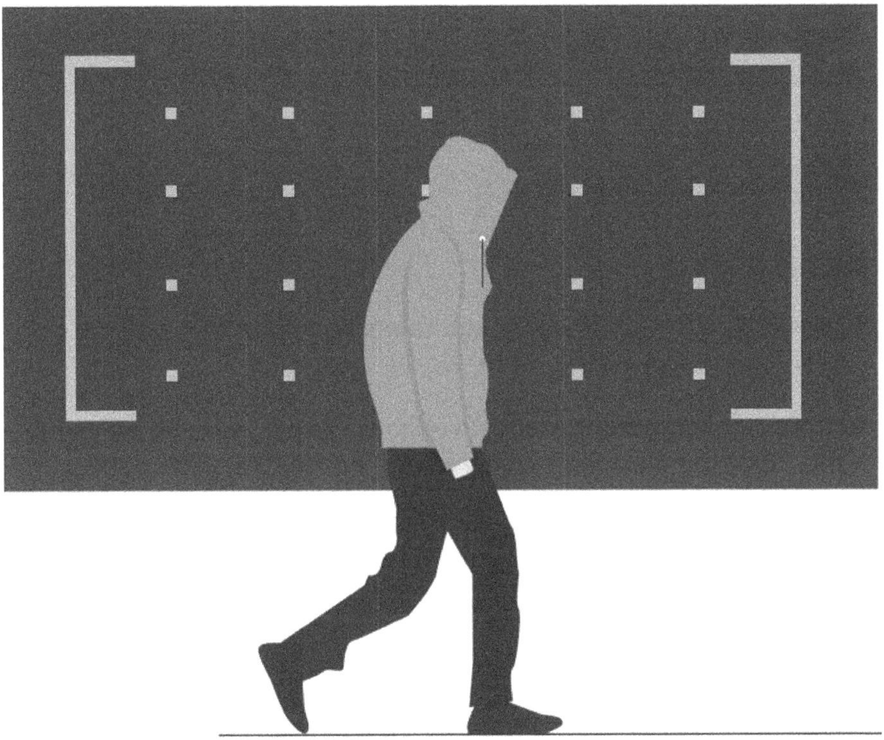

source: https://r-coder.com/matrix-r/

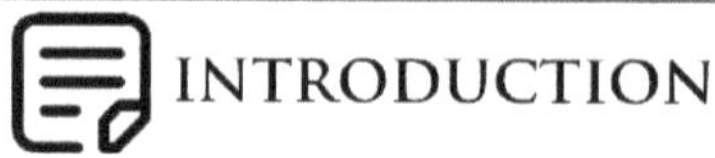

The main subject of matrices has had its origin in various linear problems; the most important concerns the nature of solutions of any given system of linear transformations in geometry. Today the subject of matrices is one of the most important and powerful tools in mathematics, which has found application to a very large numberof disciplines such as engineering, economics, statistics, physics, chemistry, biology, etc. the theory of matrices is extensively used in the solution of applied business andindustrial problems.

DISCUSSION

Definitions

A matrix is a set or group of numbers arranged in a square or rectangular arrayenclosed by two brackets.[9]

A matrix is denoted by a bold capital letter, and the elements within the matrix aredenoted by lower case letters.
e.g., matrix [**A**] with elements a_{ij} [9]

$$A_{m\times n} = mA^n$$

$$\begin{bmatrix} a_{11} & a_{12} \ldots & a_{ij} & a_{in} \\ a_{21} & a_{22} \ldots & a_{ij} & a_{2n} \\ \vdots & \vdots & \vdots & \vdots \\ a_{m1} & a_{m2} & a_{ij} & a_{mn} \end{bmatrix}$$

- i goes from 1 to m.
- j goes from 1 to n.

Matrix addition

$$a + b = (a_1 + b_1, a_2 + b_2)$$

For example, $(1,2) + (-2,3) = (1 + (-2), 2 + 3) = (-1,5)$.

We add matrices in precisely the same way; that is, by adding corresponding entries.For example,

$$\begin{bmatrix} 1 & 0 \\ 1 & 2 \end{bmatrix} + \begin{bmatrix} 2 & -1 \\ 0 & 1 \end{bmatrix} = \begin{bmatrix} 1+2 & 0+(-1) \\ 1+0 & 2+1 \end{bmatrix} = \begin{bmatrix} 3 & -1 \\ 1 & 3 \end{bmatrix}$$

And

$$\begin{bmatrix} 2 & 1 & 5 \\ 3 & 0 & 1 \end{bmatrix} + \begin{bmatrix} -2 & 4 & 2 \\ 1 & 3 & 2 \end{bmatrix} = \begin{bmatrix} 0 & 5 & 7 \\ 4 & 3 & 3 \end{bmatrix}$$

Only matrices of the same size can be added together. For example, the following is meaningless:

$$\begin{bmatrix} 1 & 2 & 3 \\ 4 & 5 & 6 \end{bmatrix} + \begin{bmatrix} -2 & 4 & 2 \\ 1 & 3 & 2 \end{bmatrix} = \begin{bmatrix} 0 & 5 & 7 \\ 4 & 3 & 3 \end{bmatrix}$$

We now write down a general formula for the addition of two matrices.

Definition

The sum of two $m \times n$ matrices **A** = $\boldsymbol{a_{ij}}$ and **B** = $\boldsymbol{b_{ij}}$ is the $m \times n$ matrix $A + B = (\boldsymbol{a_{ij}} + \boldsymbol{b_{ij}})$ given by

$$A + B = \begin{pmatrix} a_{11} + b_{11} & a_{12} + b_{12} & a_{1n+}b_{1n} \\ a_{21} + b_{21} & a_{22} + b_{22} & a_{2n} + b_{2n} \\ a_{m1} + b_{m1} & a_{m1} + b_{m1} & a_{2n} + b_{2n} \end{pmatrix}$$

Exercise: Evaluate the following matrix sums, where possible

$$\begin{pmatrix} 1 & -3 \\ -2 & 54 \end{pmatrix} + \begin{pmatrix} 2 & 0 \\ 4 & 1 \end{pmatrix}$$

$$\begin{pmatrix} 2 & 0 \\ 4 & 1 \end{pmatrix} + \begin{pmatrix} 1 & -3 \\ -2 & 54 \end{pmatrix}$$

$$\begin{pmatrix} 1 & 2 \\ 1 & 0 \\ 4 & 1 \end{pmatrix} + \begin{pmatrix} 1 & 2 & 2 \\ 1 & 3 & 1 \\ -2 & 4 & 5 \end{pmatrix}$$

$$\begin{pmatrix} 0 & 6 & -2 \\ 1 & 8 & 2 \\ 0 & 3 & 4 \end{pmatrix} + \begin{pmatrix} 1 & 2 & 9 \\ 1 & 0 & 4 \\ 3 & -4 & 1 \end{pmatrix}$$

In Exercise, parts (a) and (b) give the same answer. The commutative law, $a + b = b + a$, holds for the addition of scalars, as does the associative law, $a + (b + c) = (a + b) + c$, for all $a, b, c \in R$. Since matrix addition is defined as adding scalars, matrix addition is both commutative and associative.

Properties of Addition

The basic properties of addition for real numbers also hold true for matrices.[10]

Let A, B and C be m x n matrices

$A + B = B + A$	*commutative*
$A + (B + C) = (A + B) + C$	*associative*

There is a unique m x n matrix O with

$A + O = A$	*additive identity*

For any m x n matrix A there is an m x n matrix B (called − A) with

$A + B = O$	*additive inverse*

Proof: We have

$(A + B)_{ij} = A_{ij} + B_{ij}$ *definition of addition of matrices*

$= B_{ij} + A_{ij}$ *commutative property of addition for real numbers*

$= (B + A)_{ij}$ *definition of addition of matrices*

Zero matrix

We have seen that matrices can be added in a way analogous to the addition of scalars. For example, there is a type of matrix that corresponds to the number 0, namely, the zero matrix, which, as its name suggests, is a matrix of 0s. It is denoted by $0_{m,n}$ or by 0 when it is clear from the context which size of the matrix is intended.

Definition The $m \times n$ zero matrix $0_{m,n}$ is the $m \times n$ matrix in which all entries are 0.

The following are all examples of zero matrices:
$\begin{pmatrix} 0 \\ 0 \end{pmatrix}, (0 \quad 0 \quad 0)$

The following exercise shows that the zero matrix acts in the same way as the number0 under the operation of addition. The zero matrix is the identity element for the operation of matrix addition.
Exercise: Show that $A + 0 = A$ for any matrix $A = A_{ij}$

Definition

The negative of an $m \times n$ *matrix* $A = A_{ij}$ *is the* $m \times n$ *matrix* $-A = (-A_{ij})$

For example, if $A=\begin{pmatrix} 1 & -2 & 3 \\ -4 & 5 & -6 \end{pmatrix}$ Then

$$-A=\begin{pmatrix} -1 & 2 & -3 \\ 4 & -5 & 6 \end{pmatrix}$$

The negative of a matrix acts in the same way as the negative of a number under the operation of addition.

$$Let\ A\ be\ a\ matrix.\ Then\ A + (-A) = (-A) + A = 0$$

Proof Let $A = Aij$ so $-A = -Aij$ We add corresponding entries: the (i, j) −entry of the $matrix\ A + (-A)\ is\ aij + (-aij) = 0$. Thus the matrix $A + (-A)$ is the $zero\ matrix$ 0. Matrix addition is commutative, so $(-A) + A = A + (-A)$. Thus $(-A) + A$ is also the zero matrix 0.

Multiplication of a Matrix by a Scalar

Multiplication of a vector by a scalar generalizes in an obvious way to matrices. To multiply the vector $a = a_1, a_2$ by the scalar k, we multiply each entry in turn by k; this gives $ka = ka_1, ka_2$ Similarly, to multiply a matrix by a scalar k, we multiply each entry by k. For example,

$$3\begin{pmatrix} 1 & 2 & 3 \\ 4 & 5 & 6 \end{pmatrix} = \begin{pmatrix} 3 & 6 & 9 \\ 12 & 15 & 18 \end{pmatrix}$$

$$-\tfrac{1}{2}\begin{pmatrix} -4 & 2 \\ 0 & -6 \end{pmatrix} = \begin{pmatrix} 2 & -1 \\ 0 & 3 \end{pmatrix}$$

Definition The scalar multiple of an $m \times n$ matrix $A = A_{ij}$ by a scalar k is $m \times n$ matrix.

$$kA = \begin{pmatrix} ka_{11} & ka_{12} & \cdots & ka_{1n} \\ ka_{21} & ka_{22} & \cdots & ka_{2n} \\ \vdots & \vdots & \ddots & \vdots \\ ka_{m1} & ka_{m2} & \cdots & ka_{mn} \end{pmatrix} = (kaij)$$

Notice that $(-1)A = -A$ *and* $0A = 0$.

Properties of Scalar Multiplication

For all matrices A and B of the same size, and all scalars k, the distributive law holds;that is,

$$k(A + B) = kA + kB$$

Let r and s be real numbers and A and B be matrices. Then[10]

(r(A + B))ij = (r)(A + B)ij	definition of scalar multiplication
= (r)(Aij + Bij)	definition of addition of matrices
= r Aij + rBij	distributive property of the real numbers
= (rA)ij + (rB)ij	definition of scalar multiplication
= (rA + rB)ij	definition of addition of matrices

Matrix Multiplication

The addition of matrices generalizes the addition of scalars and vectors. A method of 'multiplying' vectors—the dot product. Let a = (a1, a2, a3) and b = (b1, b2, b3) be two vectors in R 3. Then

$$\boldsymbol{a} \times \boldsymbol{b} = \boldsymbol{a}_1\boldsymbol{b}_1 + \boldsymbol{a}_2\boldsymbol{b}_2 + \boldsymbol{a}_3\boldsymbol{b}_3$$

Matrix multiplication is a generalization of this idea. To form the product of two matrices A *and* B, we combine the rows of A with the columns of B. The (i, j) −entry of the product AB is obtained by multiplying the entries in the ith row of A with the corresponding entries in the jth column of B, and summing the products obtained. Forexample, let.[11]

$$A = \begin{pmatrix} 1 & 2 & 3 \\ 4 & 5 & 6 \end{pmatrix} \text{ and } B = \begin{pmatrix} 1 & 2 & 3 \\ 4 & 5 & 6 \\ 7 & 8 & 9 \end{pmatrix}$$

To obtain the (1, 2) −entry of the product AB, we combine the first row of A with the second column of B: we multiply together the first entry of the first row of A and the first entry of the second column of B, then multiply together the second entries of each, and then the third entries, finally adding these three numbers. The (1, 2)-entry of AB is *thus* (1 × 2) + (2 × 5) + (3 × 8) = 2 + 10 + 24 = 36.

To obtain the (2, 3)-entry of the product AB, we combine the second row of A with the third column of B. The (2, 3)-entry of AB *is* (4 × 3) + (5 × 6) + (6 × 9) = 12 + 30+ 54 = 96. We can compare how we combined the first row of A and the second column of B with the dot product of two vectors: (1, 2, 3). (2, 5, 8) =

$$(1 \times 2) + (2 \times 5) + (3 \times 8) = 36.$$

We, therefore, say that we take the dot product of the first row of A with the second column of B to obtain the (1, 2) −entry of the product AB.

One way to remember how to multiply matrices A and B is to picture running along the rows of A and then diving down the columns of B.[12]

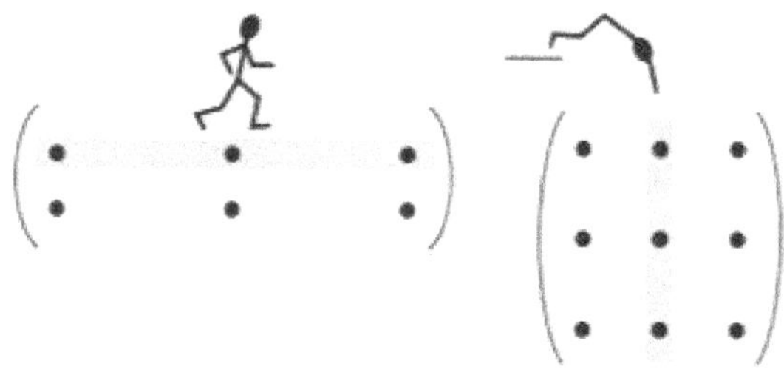

Likewise, we take the dot product of the second row of A with the third column of B toobtain $the\ (2,3) - entry\ of\ the\ product\ AB$: $(4, 5, 6).(3, 6, 9) = (4 \times 3) + (5 \times 6) + (6 \times 9) = 96$

Exercise: Find the (2, 1)-entry of the product $\boldsymbol{AB}$, for the matrices A *and* B given above.

To form the matrix product $\boldsymbol{AB}$, we combine each row of A with each column of B. We, therefore, obtain 2×3 entries in the product $\boldsymbol{AB}$, so this matrix has 2 rows and 3 columns. We have found three of the six entries, as highlighted below.

$$\begin{pmatrix} 1 & 2 & 3 \\ 4 & 5 & 6 \end{pmatrix}\begin{pmatrix} 1 & 2 & 3 \\ 4 & 5 & 6 \\ 7 & 8 & 9 \end{pmatrix} = \begin{pmatrix} 30 & 36 & 42 \\ 66 & 81 & 96 \end{pmatrix}$$

Notice that we can form the product $\boldsymbol{AB}$ since the matrix A has the same number of entries in a row as the matrix B has in a column; that is, the number of columns of A is equal to the number of rows of B. The product AB is not defined when the number of columns of the matrix A is not equal to the number of rows of the matrix B.[13]

Definition The product of an $m \times n$ matrix A with an $n \times p$ matrix B is the $m \times p$ matrix AB whose (i, j) −entry is the dot product of the ith row of A with the jth column B. [14]

(Notice that the product AB has the same number of rows as the matrix A, and thesame number of columns as the matrix B.

Schematically.

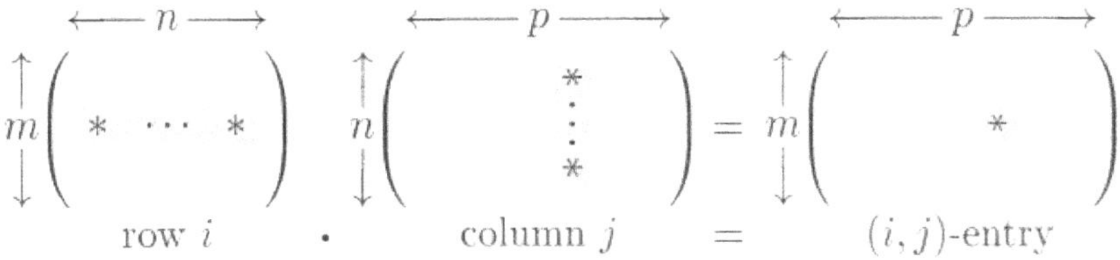

$$\overset{\longleftarrow n \longrightarrow}{m\updownarrow\begin{pmatrix} * & \cdots & * \end{pmatrix}}\ \overset{\longleftarrow p \longrightarrow}{n\updownarrow\begin{pmatrix} * \\ \vdots \\ * \end{pmatrix}} = \overset{\longleftarrow p \longrightarrow}{m\updownarrow\begin{pmatrix} & * & \end{pmatrix}}$$

row i $\cdot$ column j $=$ (i, j)-entry

Example: Evaluate (where possible) the matrix products AB, where:

$$A = \begin{pmatrix} 2 & 1 \\ -3 & 0 \end{pmatrix} \text{ and } B = \begin{pmatrix} 3 & -2 & 0 \\ 1 & 1 & 4 \end{pmatrix}$$

Solution

Matrix A has two columns, and matrix B has two rows, so the product $\boldsymbol{AB}$ can beformed. Since A has two rows and B has three columns, AB has 2 rows and three columns.

To find the (1, 1) −entry of the product $\boldsymbol{AB}$, we take the dot product of the first row of A with the first column of B: $(2 \times 3) + (1 \times 1) =$

7. Next, to find the (2, 1) −entry of AB, we take the dot product of the second row of A with the first column of B: $(-3 \times 3) + (0 \times 1) = -9$

Together, these give the first column of the product AB:

$$\begin{pmatrix} 7 & * & * \\ -9 & * & * \end{pmatrix}$$

To find *the* (1, 2) −entry of the product AB, we take the dot product of the first row of $\boldsymbol{A}$ with the second column of $\boldsymbol{B}$: $(2 \times -2) + (1 \times 1) = -3$. Next, to find the (2, 2)-entry of $\boldsymbol{AB}$, we take the dot product of the second row of A with the second column of B: $(-3 \times -2) + (0 \times 1) = 6$. We now have the second column of the product $\boldsymbol{AB}$:

$$\begin{pmatrix} 7 & -3 & * \\ -9 & 6 & * \end{pmatrix}$$

To find the (1, 3) −entry of the product AB, we take the dot product of the first row of A with the third column of B: $(2 \times 0) + (1 \times 4) = 4$. Next, to find the (2, 3)-entry of AB, we take the dot product of the second row of A with the third column *of* B: $(-3 \times 0) + (0 \times 4) = 0$. We now have the third column of the product $\boldsymbol{AB}$:

$$\begin{pmatrix} 7 & -3 & 4 \\ -9 & 6 & 0 \end{pmatrix}$$

Thus

$$\begin{pmatrix} 2 & 1 \\ -3 & 0 \end{pmatrix}\begin{pmatrix} 3 & -2 & 0 \\ 1 & 1 & 4 \end{pmatrix} = \begin{pmatrix} 7 & -3 & 4 \\ -9 & 6 & 0 \end{pmatrix}$$

Note: (When evaluating a product of matrices, it is advisable to find the entries systematically, either column by column, or row by row. Here, we find the entries column by column)

Properties of Matrix Multiplication

Unlike matrix addition, the properties of the multiplication of real numbers do not all generalize to matrices. Matrices rarely commute, even if AB and BA are both defined. There often is no multiplicative inverse of a matrix, even if the matrix is square. A few properties of multiplication of real numbers generalize to matrices. We state them now.[10]

Let A, B and C be matrices of dimensions such that the following are defined. Then

$A(BC) = (AB)C$	*associative*
$A(B + C) = AB + AC$	*distributive*
$(A + B)C = AC + BC$	*distributive*

There are unique matrices Im and In with

$ImA = AIn$	*multiplicative identity*

Proof of Property 2

Again, we show that the general element of the left-hand side is the same as the right-hand side. We have

(A(B + C))ij = S(Aik(B + C)kj)	definition of matrix multiplication
= S (Aik(Bkj + Ckj))	definition of matrix addition
= S (AikBkj + AikCkj)	distributive property of the real numbers
= S AikBkj + S AikCkj	commutative property of the real numbers
= (AB)ij + (AC)ij	definition of matrix multiplicationwhere the sum is taken from 1 to k.

Transposition of matrices

There is a simple operation that we can perform on matrices. This operation, called transposition or taking the transpose, entails interchanging the rows with the columns of the matrix. Thus, the transpose of the matrix **A,** denoted by A^T, has the rows of **A** as its columns, taken in the same order. For example,

$$\begin{pmatrix} 1 & 2 & 3 \\ 4 & 5 & 6 \\ 7 & 8 & 9 \end{pmatrix}^T = \begin{pmatrix} 1 & 4 & 7 \\ 2 & 5 & 8 \\ 3 & 6 & 9 \end{pmatrix}$$

And $$\begin{pmatrix} 2 & 7 \\ -6 & 1 \\ 0 & 4 \end{pmatrix}^T = \begin{pmatrix} 2 & -6 & 0 \\ 7 & 1 & 4 \end{pmatrix}$$

Definition The transpose of an $m \times n$ matrix **A** is the n × m matrix AT whose (i, j) −entry is the (j, i) −entry of **A.**

Exercise:

$$\begin{pmatrix} 1 & 4 \\ 0 & 2 \\ -6 & 10 \end{pmatrix} \qquad \begin{pmatrix} 2 & 1 & 2 \\ 0 & 3 & -5 \\ 4 & 7 & 0 \end{pmatrix}$$

The identity matrix I is not changed by taking the transpose; that is, $I^T = I$. The rows of matrix A forms the columns of matrix A^T , and the columns of A^T form the rows of $(A^T)^T$. Therefore the rows of A form the rows of $(A^T)^T$; that is, these two matrices are equal: $(A^T)^T = A$

Exercise let

A= $\begin{pmatrix} 1 & 2 \\ 3 & 4 \\ 5 & 6 \end{pmatrix}$ $B = \begin{pmatrix} 7 & 8 \\ 9 & 10 \\ 11 & 12 \end{pmatrix}$ $and\ C = \begin{pmatrix} 1 & 1 \end{pmatrix}$

Find A^T ,B^T and $(A + B)^T$ and verify that $(A + B)^T = A^T + B^T$. Find C^T and $(AC)^T$, and discover an equation relating $(AC)^T$,$A^T and\ C^T$

Properties of the Transpose of a Matrix

Recall that the transpose of a matrix is the operation of switching rows and columns. We state the following properties. We proved the first property in the last section.[10]

Let r be a real number and A and B be matrices. Then

$(AT)^T = A$

$(A + B)^T = A^T + B^T$ $(AB)^T = B^TA^T$

$(rA)^T = rA^T$

1. $(\mathbf{A+B})^T = \mathbf{A}^T + \mathbf{B}^T$

$$\begin{vmatrix} 7 & 3 & -1 \\ ? & -5 & 6 \end{vmatrix} + \begin{vmatrix} 1 & 5 & 6 \\ -4 & -2 & 3 \end{vmatrix} = \begin{vmatrix} 8 & 8 & 5 \\ -2 & -7 & 9 \end{vmatrix} \rightarrow \begin{vmatrix} 8 & -2 \\ 8 & -7 \\ 5 & 9 \end{vmatrix}$$

$$\begin{bmatrix} 7 & 2 \\ 3 & -5 \\ -1 & 6 \end{bmatrix} + \begin{bmatrix} 1 & -4 \\ 5 & -2 \\ 6 & 3 \end{bmatrix} = \begin{bmatrix} 8 & -2 \\ 8 & -7 \\ 5 & 9 \end{bmatrix}$$

2. $(\mathbf{AB})^T = \mathbf{B}^T\,\mathbf{A}^T$

$$\begin{vmatrix} 1 & 1 & 0 \\ ? & 2 & 3 \end{vmatrix} \begin{vmatrix} 1 \\ 1 \\ 2 \end{vmatrix} = \begin{vmatrix} 2 \\ 8 \end{vmatrix} \rightarrow |2 \quad 8|$$

$$[1 \quad 1 \quad 2] \begin{bmatrix} 1 & 0 \\ 1 & 2 \\ 0 & 3 \end{bmatrix} = [2 \quad 8]$$

SPECIAL TYPES OF MATRICES

1. Column matrix:

The number of rows may be any integer, but the number of columns is always 1.[15]

$$\begin{bmatrix}1\\4\\2\end{bmatrix}\begin{bmatrix}1\\-3\end{bmatrix}\begin{bmatrix}a_{11}\\a_{21}\\\vdots\\a_{m1}\end{bmatrix}$$

2. Row matrix:

Any number of columns but only one row

$$\begin{bmatrix}1 & 1 & 6\end{bmatrix}\quad\begin{bmatrix}0 & 3 & 5 & 2\end{bmatrix}$$

$$\begin{bmatrix}a_{11} & a_{12} & a_{13}\cdots & a_{1n}\end{bmatrix}$$

3. Rectangular matrix:

Contains more than one element and number of rows is not equal to the number ofcolumns.[15]

$$\begin{bmatrix}1 & 1\\3 & 7\\7 & -7\\7 & 6\end{bmatrix}\begin{bmatrix}1 & 1 & 1 & 0 & 0\\2 & 0 & 3 & 3 & 0\end{bmatrix}$$

$$i.e. m \neq n$$

4. Square matrix:

The number of rows is equal to the number of columns.(a square matrix **A** has an order of m)

$$\begin{vmatrix}1 & 1\\3 & 0\end{vmatrix}\qquad\begin{vmatrix}1 & 1 & 1\\9 & 9 & 0\\6 & 6 & 1\end{vmatrix}$$

The principal or main diagonal of a square matrix is composed of all elements

$a_{ij}\ for\ whicj\ i = j$

5. **Diagonal matrix**

A square matrix where all the elements are zero except those on the main diagonal.

$$\begin{bmatrix} 1 & 0 & 0 \\ 0 & 2 & 0 \\ 0 & 0 & 1 \end{bmatrix} \quad \begin{bmatrix} 3 & 0 & 0 & 0 \\ 0 & 3 & 0 & 0 \\ 0 & 0 & 5 & 0 \\ 0 & 0 & 0 & 9 \end{bmatrix}$$

$i.e\ aij = 0 \quad for\ all\ i \neq j \quad aij \neq 0\ for\ all\ i = j$

6. **Unit or Identity matrix – I:**

A diagonal matrix with ones on the main diagonal

$$\begin{vmatrix} 1 & 0 & 0 & 0 \\ 0 & 1 & 0 & 0 \\ 0 & 0 & 1 & 0 \\ 0 & 0 & 0 & 1 \end{vmatrix} \qquad \begin{vmatrix} 1 & 0 \\ 0 & 1 \end{vmatrix} \begin{vmatrix} a_{ij} & 0 \\ 0 & a_{ij} \end{vmatrix}$$

$i.e. a_{ij} = 0\ for\ all\ i = j \qquad a_{ij} = 1\ for\ some\ or\ all\ i = j$

7. **Null (zero) matrix – 0:**

All elements in the matrix are zero.

$$\begin{bmatrix} 0 \\ 0 \\ 0 \end{bmatrix} \begin{bmatrix} 0 & 0 & 0 \\ 0 & 0 & 0 \\ 0 & 0 & 0 \end{bmatrix}$$

$\text{i.e. } a_{ij} = 0 \text{ for all i, j}$

8. **Triangular matrix:**

A square matrix whose elements above or below the main diagonal are all zero.

$$\begin{bmatrix} 1 & 0 & 0 \\ 2 & 1 & 0 \\ 5 & 2 & 3 \end{bmatrix} \begin{bmatrix} 1 & 0 & 0 \\ 2 & 1 & 0 \\ 5 & 2 & 3 \end{bmatrix} \begin{bmatrix} 1 & 8 & 9 \\ 0 & 1 & 6 \\ 0 & 0 & 3 \end{bmatrix}$$

a. **Upper triangular matrix:**

$$\begin{bmatrix} a_{ij} & a_{ij} & a_j \\ 0 & a_{ij} & a_{ij} \\ 0 & 0 & a_{ij} \end{bmatrix} \begin{bmatrix} 1 & 8 & 7 \\ 0 & 1 & 8 \\ 0 & 0 & 3 \end{bmatrix} \begin{bmatrix} 1 & 7 & 4 & 4 \\ 0 & 1 & 7 & 4 \\ 0 & 0 & 7 & 8 \\ 0 & 0 & 0 & 3 \end{bmatrix}$$

$$i.e\ aij = 0\ for\ all\ i\ > j$$

b. **Lower triangular matrix:**

A square matrix whose elements above the main diagonal are all zero.

$$\begin{bmatrix} a_{ij} & 0 & 0 \\ a_{ij} & a_{ij} & 0 \\ a_{ij} & a_{ij} & a_{ij} \end{bmatrix} \begin{bmatrix} 1 & 0 & 0 \\ 2 & 1 & 0 \\ 5 & 2 & 3 \end{bmatrix}$$

$$i.e. aij\ = 0\ for\ all\ i < j$$

9. **Scalar matrix:**
A diagonal matrix whose main diagonal elements are equal to the same scalar.A scalar is defined as a single number or constant.[15]

$$\begin{bmatrix} a_{ij} & 0 & 0 \\ 0 & a_{ij} & 0 \\ 0 & 0 & a_{ij} \end{bmatrix} \begin{bmatrix} 1 & 0 & 0 \\ 0 & 1 & 0 \\ 0 & 0 & 1 \end{bmatrix} \begin{bmatrix} 6 & 0 & 0 & 0 \\ 0 & 6 & 0 & 0 \\ 0 & 0 & 6 & 0 \\ 0 & 0 & 0 & 6 \end{bmatrix}$$

$$i.e. aij = 0\ for\ all\ i = j \qquad aij\ = a\ for\ all\ i = j$$

EXERCISE 1.2

1. $A = \begin{pmatrix} 5 & 2 \\ 0 & 1 \\ 1 & 9 \end{pmatrix}$ $B = \begin{pmatrix} 2 & 3 \\ 4 & 1 \\ 0 & 2 \end{pmatrix}$ $A + B = ?$

2. $C = \begin{pmatrix} 3 & 4 & 9 \\ 6 & 8 & 6 \\ 7 & 3 & 4 \end{pmatrix}$ $D = \begin{pmatrix} 1 & 6 & 7 \\ 6 & 4 & 2 \\ 4 & 1 & 5 \end{pmatrix}$ $C - D = ?$

3. Let $A = \begin{pmatrix} 5 & -3 \\ 2 & 3 \\ -1 & 0 \end{pmatrix}$ $and\ B = \begin{pmatrix} 2 & 1 \\ -2 & -7 \\ 3 & 5 \end{pmatrix}$

Evaluate the following:

(a) $4A$

(b) $4B$

(c) $4A + 4B$

(d) $4(A + B)$

4. Evaluate the following matrix products,

A. $\begin{pmatrix} 2 & -1 \\ 0 & 3 \\ 1 & 2 \end{pmatrix} \begin{pmatrix} 3 \\ 2 \end{pmatrix}$

$\begin{pmatrix} 3 & 1 & 2 \\ 0 & 5 & 1 \end{pmatrix} \begin{pmatrix} -2 & 0 & 1 \\ 1 & 3 & 0 \\ 4 & 1 & -1 \end{pmatrix}$

5. X= A + 3B

$$A = \begin{pmatrix} 1 & 5 & -1 \\ -1 & 2 & 2 \\ 0 & 3 & 3 \end{pmatrix} \ \ and \ \ B = \begin{pmatrix} -1 & -4 & 3 \\ 1 & -2 & -2 \\ -3 & 3 & -5 \end{pmatrix}$$

1. Evaluate the following matrix difference:

a. $\begin{pmatrix} 3 & 0 \\ 2 & 7 \end{pmatrix} - \begin{pmatrix} 10 & 3 \\ 1 & 5 \\ 15 & 12 \end{pmatrix}$

b. $\begin{pmatrix} 5 & 8 & 12 \\ 7 & 2 & -1 \end{pmatrix} - \begin{pmatrix} 3 & 10 & 2 \\ 4 & 9 & 21 \end{pmatrix}$

2. Evalaute (where possible) the matrix products AB, where:

$$A = \begin{pmatrix} 2 & 1 \\ -3 & 0 \end{pmatrix} \ and\ B = \begin{pmatrix} 3 & -2 \end{pmatrix}$$

3. State which matrices are diagonal, upper-triangular, or lower-triangular.

$a.\begin{pmatrix} 1 & 1 & 1 \\ 0 & 2 & 2 \\ 0 & 0 & 3 \end{pmatrix}$

$b.\ \begin{pmatrix} 9 & 0 \\ 0 & 0 \end{pmatrix}$

$c.\begin{pmatrix} 0 & 0 & 1 \\ 0 & 1 & 2 \\ 1 & 2 & 3 \end{pmatrix}$

$d.\begin{pmatrix} 1 & 0 \\ 1 & 0 \end{pmatrix}$

4. Write down the transpose of the given matrix

$$\begin{pmatrix} 1 & 4 \\ 0 & 2 \\ -6 & 10 \end{pmatrix} \begin{pmatrix} 2 & 1 & 2 \\ 0 & 3 & -5 \\ 4 & 7 & 0 \end{pmatrix}$$

5. Suppose that we are given matrices of the following sizes:

$A : 2 \times 1$

$B : 4 \times 3$

$C : 3 \times 2$

$D : 1 \times 4$

E : 3 X 3

F : 3 X 4

G : 2 X 4

Which of the following expressions are defined? Give the size of the resultingmatrix for those that are defined:

a. $FB + E$

b. $GF^T - ADB$

c. $BF - (FB)^T$

d. $C(AD + G)$

6. Let $A = \begin{pmatrix} 1 & 6 \\ 3 & -4 \end{pmatrix}$ *and* $B = \begin{pmatrix} 2 & -3 \\ 0 & 7 \end{pmatrix}$. Evaluate the following:

a. BA
b. A^2
c. $A^T B^T$

KEY TO CORRECTION

Exercises (1.1)	Exercises (1.2)
1. (x = 2, y = 3.) **2.** (x = 5, y = 6 and z = −3) **3.** (p = 7 and q = 3) **4.** (x = 2, y = 1) **5.** (4, -1) **6.** (2, -1) **7.** $(3,\frac{1}{2})$ **8.** The answer is $(2,8)$ **9.** $(3,-4)$ **10.** $(4,2)$	1. $\begin{pmatrix} 7 & 5 \\ 4 & 2 \\ 1 & 11 \end{pmatrix}$ 2. $\begin{pmatrix} 2 & -2 & 2 \\ 0 & 4 & 4 \\ 3 & 2 & -1 \end{pmatrix}$ 3. $a.\begin{pmatrix} 20 & -12 \\ 8 & 12 \\ -4 & 0 \end{pmatrix} b.\begin{pmatrix} 8 & 4 \\ -8 & -28 \\ 12 & 20 \end{pmatrix}$ $c.\begin{pmatrix} 28 & -8 \\ 0 & -16 \\ 8 & 20 \end{pmatrix} d.\begin{pmatrix} 28 & -8 \\ 0 & -16 \\ 8 & 20 \end{pmatrix}$ 4. A. $\begin{pmatrix} 4 \\ 6 \\ 7 \end{pmatrix}$ B. $\begin{pmatrix} 3 & 5 & 1 \\ 9 & 16 & -1 \end{pmatrix}$ 5. $x = \begin{pmatrix} 2 & -7 & 8 \\ 2 & -4 & -4 \\ -9 & 6 & 12 \end{pmatrix}$

Chapter 2

SOLVING LINEAR SYSTEMS

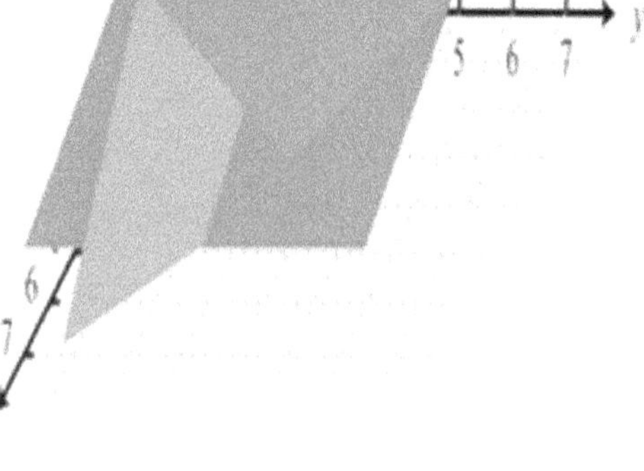

source: Saylordotorg.github.io

SOLVING LINEAR SYSTEMS

After working through this section, you should be able to:

- Solve the echelon form of a matrix.
- Perform elementary row-echelon of a matrix.
- Solve elementary matrices in finding inverse matrices.
- Solve the linear system by Gaussian elimination and Gauss Jordan Reduction.
- Solve linear system in reduced row echelon form.
- Apply elementary row operations to solve linear systems of equations.

Lesson 2.1 Echelon Form of a Matrix

INTRODUCTION

In lesson 1 we discussed the method of elimination for solving linear systems, coefficient matrix, and augmented matrix, which, when applied to an augmented matrix, can greatly simplify the steps needed to determine the solution of the associated linear system. The operations discussed in this section apply to any matrix, whether it is an augmented matrix. in 2.2, we apply the construction developed in this section to the solution of the linear system.

Using the augmented matrix of a linear system and an echelon form, we develop two methods for solving a system of m linear equations in n unknowns. These methods take the augmented matrix of the linear system, perform elementary row operations on it, and obtain a new matrix representing an equivalent linear system (i.e., a system with the same solutions as the original linear system). The important point is that the latter linear system can be solved more easily.[1]

DISCUSSION

We begin by defining the various elements that make up a matrix.

DEFINITION 2.1

An m x n matrix is a rectangular array of numbers with m rows and n columns.[2]

$$\underbrace{\begin{bmatrix} a_{11} & a_{12} & a_{13} & \cdots & a_{1n} \\ a_{21} & a_{22} & a_{23} & \cdots & a_{2n} \\ a_{31} & a_{32} & a_{33} & \cdots & a_{3n} \\ \vdots & \vdots & \vdots & \ddots & \vdots \\ a_{m1} & a_{m2} & a_{m3} & \cdots & a_{mn} \end{bmatrix} \leftarrow \left.\vphantom{\begin{matrix}a\\a\\a\\a\\a\end{matrix}}\right\}}_{n\ columns} m\ rows$$

We say the matrix has dimension *m* x *n*. The numbers a_{ij} are the entries of the matrix.[2]

- The subscript on the entry a_{ij} indicates that it is in the *ith* row and the *j*th column.

A m x n matrix A is said to be in reduced row-echelon form if it satisfies the followingproperties:

a. The first nonzero entry from the left of a nonzero row is a 1. This entry is called a leading one of its rows.[1]

b. For each nonzero row, the leading one appears to the right and below anyleading ones in preceding rows.

c. If there are any, all zero rows appear at the bottom of the matrix.[1]

d. If a column contains a leading one, then all other entries in that column are
zero.[1]

A matrix in a reduced row-echelon appears like a staircase (''echelon'') pattern of leading ones descending from the upper left corner of the matrix. An m x n matrix satisfying properties (a), (b), and (c) is said to be in row echelon form. In Definition 2.1, there may be no zero rows. A similar definition can be formulated obviously for reduced column echelon form and column echelon form.[1]

In the following matrices, the first is not in row-echelon form. The second is in row-echelonform. The third is in reduced row-echelon form. The entries in red are the leading entries.[2]

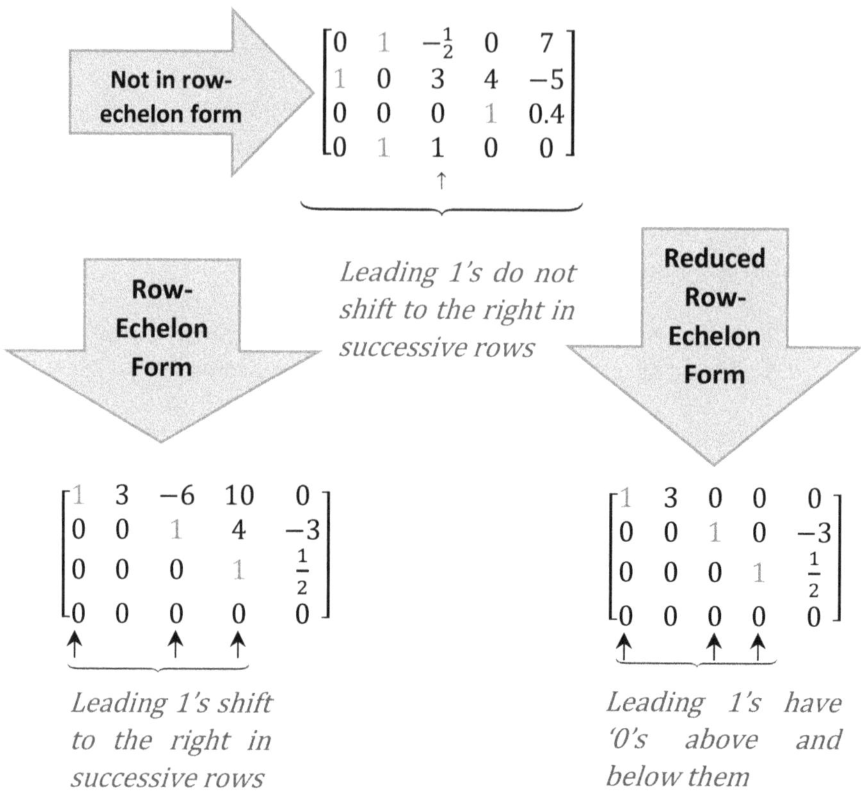

We now discuss a systematic way to use elementary row operations to put a matrix in row-echelon form. We see how the process might work for a 3 x 4 matrix.[3]

Step 1: Start by obtaining 1 in the top left corner. Then, obtain zeros below that 1 by adding appropriate multiples of the first row to the rows below it.[2]

$$\begin{bmatrix} 1 & \_ & \_ & \_ \\ 0 & \_ & \_ & \_ \\ 0 & \_ & \_ & \_ \end{bmatrix}$$

STEP 2: Next, obtain a leading 1 in the next row

STEP 3: Then, obtain zeros below that 1

- At each stage, make sure every leading entry is to the right of the leadingentry in the row above it.

- Rearrange the rows if necessary.

$$\begin{bmatrix} 1 & \_ & \_ & \_ \\ 0 & \_ & \_ & \_ \\ 0 & \_ & \_ & \_ \end{bmatrix} \quad \begin{bmatrix} 1 & \_ & \_ & \_ \\ 0 & 1 & \_ & \_ \\ 0 & 0 & \_ & \_ \end{bmatrix}$$

STEP 4: Continue this process until you arrive at a matrix in row-echelon form

$$\begin{bmatrix} 1 & \_ & \_ & \_ \\ 0 & \_ & \_ & \_ \\ 0 & \_ & \_ & \_ \end{bmatrix} \rightarrow \begin{bmatrix} 1 & \_ & \_ & \_ \\ 0 & 1 & \_ & \_ \\ 0 & 0 & \_ & \_ \end{bmatrix} \rightarrow \begin{bmatrix} 1 & \_ & \_ & \_ \\ 0 & 1 & \_ & \_ \\ 0 & 0 & 1 & \_ \end{bmatrix}$$

Once an augmented matrix is in row-echelon form, we can use back-substitution to the corresponding linear system. This technique is called Gaussian elimination, in honor of its inventor, the German mathematician Carl Friedrich Gauss.[4]

Example 1.

The following are matrices in reduced row-echelon form since they satisfy properties (a),(b), (c), and (d):

$$A = \begin{vmatrix} 1 & 2 & 0 & 0 & 1 \\ 0 & 0 & 1 & 2 & 3 \\ 0 & 0 & 0 & 0 & 0 \end{vmatrix} \qquad B = \begin{vmatrix} 1 & 0 & 0 \\ 0 & 1 & 0 \\ 0 & 0 & 1 \end{vmatrix}$$

The matrices that follow are not in reduced row-echelon form

$$C = \begin{vmatrix} 1 & 2 & 0 & 4 \\ 0 & 0 & 0 & 0 \\ 0 & 0 & 1 & -3 \end{vmatrix} \qquad D = \begin{vmatrix} 1 & 0 & 3 & 4 \\ 0 & 2 & -2 & 5 \\ 0 & 0 & 1 & 2 \end{vmatrix}$$

Example 2.

The following are matrices in row echelon form:

$$E = \begin{bmatrix} 1 & 5 & 0 & 2 & -2 & 4 \\ 0 & 1 & 0 & 3 & 4 & 8 \\ 0 & 0 & 0 & 1 & 7 & -2 \\ 0 & 0 & 0 & 0 & 0 & 0 \\ 0 & 0 & 0 & 0 & 0 & 0 \end{bmatrix} \quad F = \begin{bmatrix} 1 & 0 & 0 & 0 \\ 0 & 1 & 0 & 0 \\ 0 & 0 & 1 & 0 \\ 0 & 0 & 0 & 1 \end{bmatrix}$$

$$G = \begin{bmatrix} 0 & 0 & 1 & 3 & 5 & 7 & 9 \\ 0 & 0 & 0 & 0 & 1 & -2 & 3 \\ 0 & 0 & 0 & 0 & 0 & 1 & 2 \\ 0 & 0 & 0 & 0 & 0 & 0 & 1 \\ 0 & 0 & 0 & 0 & 0 & 0 & 0 \end{bmatrix}$$

This shows that every matrix can be put into row (column) echelon form, or into reduced row (column) echelon form, through certain row (column) operations.

DEFINITION 2.2

An elementary row (column) operation on a matrix A is any one of the following operations:[5]

(a). Type I: Interchange any two rows (columns):

(b). Type II: Multiply a row (column) by a nonzero number.

(c). Type III. Add a multiple of one row (column) to another.

We now introduce the following notation for elementary row and elementary column operations on matrices:[6]

- Interchange rows (columns) *i* , and *j*, Type I:

  $r_i \leftrightarrow r_j \quad (c_i \leftrightarrow c_j).$

- Replace row (column) *I* by *k* times row (column) *i*, Type II:

  $kr_i \rightarrow r_i \; (kc_i \rightarrow c_i)$

- Replace row (column) *j* by *k* times row (column) *i+ row* (column) *j*, Type III:

  $kr_j + r_j \rightarrow rj(kc_j + c_j \rightarrow cj)$

Example 3.

Let $A = \begin{vmatrix} 0 & 0 & 1 & 2 \\ 2 & 3 & 0 & -2 \end{vmatrix}$

Interchanging rows 1 and 3 of A, we obtain

$$B = \begin{vmatrix} 3 & 3 & 6 & -9 \\ 2 & 3 & 0 & -2 \\ 0 & 0 & 1 & 2 \end{vmatrix}$$

Multiplying the third row of A by $\frac{1}{3}$, $A = \frac{1}{3}\begin{vmatrix} 0 & 0 & 1 & 2 \\ 2 & 3 & 0 & -2 \\ \mathbf{3} & \mathbf{3} & \mathbf{6} & \mathbf{-9} \end{vmatrix}$ we obtain

$$C = \begin{vmatrix} 0 & 0 & 1 & 2 \\ 2 & 3 & 0 & -2 \\ 1 & 2 & 2 & -3 \end{vmatrix}$$

Adding (-2) times row 2 of A to row 3 of A

$A = -2\begin{vmatrix} 0 & 0 & 1 & 2 \\ 2 & 3 & 0 & -2 \\ 3 & 3 & 6 & -9 \end{vmatrix}$ we obtain $D = \begin{vmatrix} 0 & 0 & 1 & 2 \\ 2 & 3 & 0 & -2 \\ -1 & -3 & 6 & -5 \end{vmatrix}$

Observe that in obtaining D from A, row 2 of A did not change.

DEFINITION 2.3

An m x n matrix B is said to be row (column) equivalent to an m x n matrix A if B can be produced by applying a finite sequence of elementary row (column) operations to A.

Example 4.

Let $A = \begin{vmatrix} 1 & 2 & 4 & 3 \\ 2 & 1 & 3 & 2 \\ 1 & -2 & 2 & 3 \end{vmatrix}$

If we add 2 times row 2 of A to its third row, we obtain.

$B = \begin{vmatrix} 1 & 2 & 4 & 3 \\ 2 & 1 & 3 & 2 \\ 5 & 0 & 8 & 7 \end{vmatrix}$ so B is row equivalent to A. Interchanging rows 1 and 3 of B, we obtain $C = \begin{vmatrix} 5 & 0 & 8 & 7 \\ 2 & 1 & 3 & 2 \\ 2 & 2 & 4 & 3 \end{vmatrix}$ so C is row equivalent to B. Multiply row 3 of C by 2, we obtain $D = \begin{vmatrix} 5 & 0 & 8 & 7 \\ 2 & 1 & 3 & 2 \\ 2 & 4 & 8 & 6 \end{vmatrix}$ so D is row equivalent to B. Multiplying row 3 of C by 2, we obtain so D is row equivalent to C. It then follows that D is row equivalent to A since we obtained D by applying three successive elementary row operations to A.[6]

THEOREM 2.1. Every nonzero m x n matrix $A = [a_{ij}]$ is row (column) equivalent to a matrix in row (column) echelon form. [6]

Example 5

Let $$A = \begin{bmatrix} 0 & 2 & 3 & -4 & 1 \\ 0 & 0 & 2 & 3 & 4 \\ 2 & 2 & -5 & 2 & 4 \\ 2 & 0 & -6 & 9 & 7 \end{bmatrix}$$

Pivot column　　Pivot

Column 1 is the first (counting from left to right) column in A with a nonzero entry, so column 1 is the pivot column of A. the first (counting from top to bottom) nonzero entry in the pivot column occurs in the third row, so the pivot is $a_{31} = 2$.

We interchange the first and third rows of A, obtaining

$$B = \begin{bmatrix} 2 & 2 & -5 & 2 & 4 \\ 0 & 0 & 2 & 3 & 4 \\ 0 & 2 & 3 & -4 & 1 \\ 2 & 0 & -6 & 9 & 7 \end{bmatrix},$$

Multiply the first row of B by the reciprocal of the pivot, that is, by

$\frac{1}{b_{11}} = \frac{1}{2}$ to obtain $C = \begin{bmatrix} 1 & 1 & \frac{-5}{2} & 1 & 2 \\ 0 & 0 & 2 & 3 & 4 \\ 0 & 2 & 3 & -4 & 1 \\ 2 & 0 & -6 & 9 & 7 \end{bmatrix}$

Add (-2) times the first row of C to the fourth row of C to produce a matrix D in which theonly nonzero entry in the pivot column is $d_{11} = 1$:

$$D = \begin{bmatrix} 1 & 1 & \frac{-5}{2} & 1 & 2 \\ 0 & 0 & 2 & 3 & 4 \\ 0 & 2 & 3 & -4 & 1 \\ 0 & -2 & -1 & 7 & 3 \end{bmatrix}$$

Interchange the second row and third row of D, obtaining

$$E = \begin{bmatrix} 1 & 1 & \frac{-5}{2} & 1 & 2 \\ 0 & 2 & 3 & -4 & 1 \\ 0 & 0 & 2 & 3 & 4 \\ 0 & -2 & -1 & 7 & 3 \end{bmatrix}$$

Multiply the second row of second column of E by the reciprocal of the pivot, that is, by $^1/_2$ to obtain

$$F = \begin{bmatrix} 1 & 1 & \frac{-5}{2} & 1 & 2 \\ 0 & 1 & \frac{3}{2} & -2 & \frac{1}{2} \\ 0 & 0 & 2 & 3 & 4 \\ 0 & -2 & -1 & 7 & 3 \end{bmatrix}$$

Multiply 2 to the second row then add to fourth row of the pivot column to obtain

$$G = \begin{bmatrix} 1 & 1 & \frac{-5}{2} & 1 & 2 \\ 0 & 1 & & & \\ 0 & 0 & & & \\ & & & & \end{bmatrix}$$

Multiply the third row of third column of G by the reciprocal of the pivot, that is, by $\frac{}{2}$ to obtain

$$H = \begin{bmatrix} & & & & \\ & & & & \\ 0 & 0 & 1 & & 2 \\ 0 & 0 & 2 & 3 & 4 \end{bmatrix}$$

Multiply -2 to the third row then add to fourth row of the pivot column to obtain

$$I = \begin{bmatrix} 1 & 1 & \frac{-5}{2} & 1 & 2 \\ 0 & 1 & \frac{3}{2} & -2 & \frac{1}{2} \\ 0 & 0 & 1 & \frac{3}{2} & 2 \\ 0 & 0 & 0 & 0 & 0 \end{bmatrix}$$

The matrix $I = \begin{bmatrix} 1 & 1 & \frac{-5}{2} & 1 & 2 \\ 0 & 1 & \frac{3}{2} & -2 & \frac{1}{2} \\ 0 & 0 & 1 & \frac{3}{2} & 2 \\ 0 & 0 & 0 & 0 & 0 \end{bmatrix}$ is in row echelon form and is row equivalent to A.

Lesson 2.2 Solving Linear Systems

To see how a linear system whose augmented matrix a form has can be readily solved, supposethat

$$\left[\begin{array}{ccc|c} 1 & 2 & 0 & 3 \\ 0 & 1 & 1 & 2 \\ 0 & 0 & 1 & -1 \end{array}\right]$$

represents the augmented matrix of a linear system. Then the solution is quickly found from the corresponding equations[7]

$x_1 + 2x_2 = 3$	as	$x_3 = -1$
$x_2 + x_3 = 2$		$x_2 = 2 - x_3 = 2+1=3$
$x_3 = -1$		$x_1 = 3 - 2x_2 = 3-6= -3.$

The task of this section is to manipulate the augmented matrix representing a given linear systeminto a form in which the solution can be found more easily.[7]

We now apply row operations to the solution of linear systems.

THEOREM 2.3

Let Ax = b and Cx = d be two linear systems, each of *m* equations in *n* unknowns. If the augmented matrices $[A \vdots B]$and $[C \vdots D]$, of these systems are row equivalent, then both linear systems have the same solutions.[5]

COROLLARY 2.1

If A and C are now equivalent *m x n* matrices, then the linear systems Ax = 0 and Cx = 0 are equivalent.

Gaussian elimination consists of two steps:

Step 1. The transformation of the augmented matrix $[A \vdots B]$ to the matrix $[C \vdots d]$ in row echelon form using elementary row operations.

Step 2. Solution of the linear system corresponding to the augmented matrix $[C \vdots d]$ using the back substitution. [7]

THE GAUSS -JORDAN REDUCTION procedure for solving linear system Ax=B is as follows:

Step 1. From the augmented matrix $[A \vdots B]$

Step 2. Transform the augmented matrix to reduced row echelon form by using elementary row operations.

Step 3. The linear system that corresponds to the matrix *m* reduced row echelon form that has been obtained in step 2 has the same solutions as the given linear system. For each nonzero row of the matrix in reduced row-echelon form, solve the corresponding equation for the unknown corresponding to the row's leading entry. The rows consisting entirely of zeros can be ignored since the corresponding equation will be satisfied for any values of the unknown.[8]

Example 1

$$x + 2y + 3z = 9$$

$$2x - y + z = 8$$

$$3x \quad\quad - z = 3$$

Has the augmented matrix $[A \vdots b] = \begin{bmatrix} 1 & 2 & 3 & \vdots & 9 \\ 2 & -1 & 1 & \vdots & 8 \\ 3 & 0 & -1 & \vdots & 3 \end{bmatrix}$

Transforming this matrix to row echelon form, we obtain (verify)

$$[C \vdots d] = \begin{bmatrix} 1 & 2 & 3 & \vdots & 9 \\ 0 & 1 & 1 & \vdots & 2 \\ 0 & 0 & 1 & \vdots & 3 \end{bmatrix} = \begin{bmatrix} 1 & 0 & 0 & \vdots & 2 \\ 0 & 1 & 0 & \vdots & -1 \\ 0 & 0 & 1 & \vdots & 3 \end{bmatrix} \leftarrow \text{Step 2}$$

Solutions x=2, y =-1 and z=3 ← Step 3

Using back substitution, we now have

z=**3**

y=2-z=2-3**=-1**

x=9-2y-3z=9+2-9=**2;**

thus, the solution is x=2, y=-1, z =3, which is unique.

Example 2.

Let $[C \vdots d] = \begin{vmatrix} 1 & 2 & 3 & 4 & 5 & \vdots & 6 \\ 0 & 1 & 2 & 3 & -1 & \vdots & 7 \\ 0 & 0 & 1 & 2 & 3 & \vdots & 7 \\ 0 & 0 & 0 & 1 & 2 & \vdots & 9 \end{vmatrix}$

Then

$$x_4 = 9 - 2x_5$$

$$x_3 = 7 - 2x_4 - 3x_5 = 7 - 2(9 - 2x_5) - 3x_5 = -11 + x_5$$

$$x_2 = 7 - 2x_3 - 3x_4 + x_5 = 2 + 5x_5$$

$$x_1 = 6 - 2x_2 - 3x_3 - 4x_4 - 5x_5 = -1 - 10x_5$$

$$x_5 = \text{any real number}$$

The system is consistent, and all solutions are of the form

$$x_1 = -1 - 10r$$

$$x_2 = 2 + 5r$$

$$x_3 = -11 + r$$

$$x_4 = 9 - 2r$$

$$x_5 = r, \text{any real number}$$

Since r can be assigned any real number, the given linear system has infinitely many solutions.

Example 3

If $[C : d] = \begin{vmatrix} 1 & 2 & 3 & 4 & \vdots & 5 \\ 0 & 1 & 2 & 3 & \vdots & 6 \\ 0 & 0 & 0 & 0 & \vdots & 1 \end{vmatrix}$, the C_x= d has no solution, since

The last equation is $0x_1 + 0x_2 + 0x_3 + 0x_4 = 1$, which can never be satisfied.

Example 4

If $[C:d]=\begin{vmatrix}1&0&0&0&:&5\\0&1&0&0&:&6\\0&0&1&0&:&7\\0&0&0&1&:&8\end{vmatrix}$, then the unique solution is $\begin{matrix}x_1&=&5\\x_2&=&6\\x_3&=&7\\x_4&=&8\end{matrix}$

Now we study a homogenous system $A_x=0$ of m linear equations in n unknowns.

The solution, $x_1=x_2=x_3=\cdots.x_n=0$ to the homogenous system $Ax=0$ Is called the TRIVIAL SOLUTION

The solution $x_1,x_2,x_3,\ldots x_n$to the homogenous system in which not all the x_i are zero is called NONTRIVIAL SOLUTION

Example 5. Consider the homogenous system

$x+2y+3z=0$

$-x+3y+2z=0$

$2x+y-2z=0$

Step 1. The augmented matrix is $\begin{vmatrix}1&2&3&:&0\\-1&3&2&:&0\\2&1&-2&:&0\end{vmatrix}$

$$\begin{vmatrix}1&0&0&:&0\\0&1&0&:&0\\0&0&1&:&0\end{vmatrix}$$

Step 3. Solution $\begin{matrix}x\rightarrow=0\\y\rightarrow=0\\z\rightarrow=0\end{matrix}$, $x=y=z=0$ therefore, it is Trivial solution

Example 5. Consider the homogenous system $\begin{matrix}x+2y+3z=0\\-x+3y+2z=0\\2x+y-2z=0\end{matrix}$

Step 1. The augmented matrix is $\begin{vmatrix}1&2&3&:&0\\-1&3&2&:&0\\2&1&-2&:&0\end{vmatrix}$

Step 2. Reduced row echelon form $\begin{vmatrix} 1 & 0 & 0 & \vdots & 0 \\ 0 & 1 & 0 & \vdots & 0 \\ 0 & 0 & 1 & \vdots & 0 \end{vmatrix}$

Step 3. Solutions $\begin{matrix} x & \rightarrow & = & 0 \\ y & \rightarrow & = & 0 \\ z & \rightarrow & = & 0 \end{matrix}$, x = y = z = 0 therefore, it is Trivial Solution

Lesson 2.3 Elementary Matrices: Finding A⁻¹

DEFINITION 2.4

An *nxn* matrix A is called nonsingular (or invertible) if an n x n matrix B exists suchthat AB=BA=In. The matrix B is called an inverse of A. if there exists no such matrix B, then A is called SINGULAR (or Nonvertible)' [9]

Example 1.

Let $A = \begin{bmatrix} -2 & 3 \\ 2 & 2 \end{bmatrix}$ $B = \begin{bmatrix} -1 & \frac{3}{2} \\ 1 & -1 \end{bmatrix}$ $AB = \begin{bmatrix} 1 & 0 \\ 0 & 1 \end{bmatrix}$

$I_2 = \begin{bmatrix} 1 & 0 \\ 0 & 1 \end{bmatrix}$ Since AB=BA=I_2 , B is an inverse of A and A is nonsingular

THEOREM 2.5

If a matrix has an inverse, then the inverse is unique.

We write the inverse of A as A^{-1}. Thus, A A^{-1} = A^{-1} A= In

Example 2

Let $A^{-1} = \begin{bmatrix} 1 & 2 \\ 3 & 4 \end{bmatrix}$ Find A^{-1}

Solution: $A^{-1} = \begin{bmatrix} a & b \\ c & d \end{bmatrix}$ $AA^{-1} = A^{-1}A = l_2$

$$\begin{bmatrix} 1 & 2 \\ 3 & 4 \end{bmatrix}\begin{bmatrix} a & b \\ c & d \end{bmatrix} = \begin{bmatrix} 1 & 0 \\ 0 & 1 \end{bmatrix} \rightarrow \begin{bmatrix} a+2c & b+2d \\ 3a+4c & 3b+4d \end{bmatrix} = \begin{bmatrix} 1 & 0 \\ 0 & 1 \end{bmatrix}$$

$a + 2c = 1$ $\quad b + 2d = 0$

$3a + 4c = 0$ $\quad 3b + 4d = 1$ using the Gaussian Elimination method, we find the values of $c = \frac{3}{2}, b = -\frac{1}{2} and\ a = 2.$

Therefore, the $A^{-1} = \begin{bmatrix} 2 & 1 \\ 3/2 & -1/2 \end{bmatrix}$ Matrix A is **nonsingular**.

Properties of the Inverse

a. If A is a nonsingular matrix, then A^{-1} is nonsingular, and $(A^{-1})^{-1} = A$.[5]
b. If A and B are nonsingular matrices, then AB is nonsingular and $(AB)^{-1} = B^{-1}A^{-1}$.[10]
c. If A is nonsingular matrix, then $(A^{T})^{-1} = (A^{-1})^{T}$

Practical Method for Finding A^{-1}

Step 1. Form the n x n matrix $[A \vdots I_n]$ obtained by adjoining the identity matrix I_n to the given matrix A.

Step 2. Transform the matrix obtained in step 1 to reduced row echelon form by using elementary row operation.[11]

Step 3. Suppose that step 2 has produced the matrix $[C \vdots D]$ in reduced row echelon form

a. If $c = I_n$, then $D=A^{-1}$
b. If $c \neq I_n$, then c has a row of zeros. In this case, A singular and A^{-1} do not exist.

Example 3. Find the inverse of the matrix, $A = \begin{vmatrix} 1 & 1 & 1 \\ 0 & 2 & 3 \\ 5 & 5 & 1 \end{vmatrix}$

Solution:

Step 1. $\begin{vmatrix} 1 & 1 & 1 & \vdots & 1 & 0 & 0 \\ 0 & 2 & 3 & \vdots & 0 & 1 & 0 \\ 5 & 5 & 1 & \vdots & 0 & 0 & 1 \end{vmatrix}$ $\qquad [C \vdots D]\ I_3 \vdots A^{-1}$

Step 2. $\begin{vmatrix} 1 & 0 & 0 & \vdots & \frac{13}{8} & -\frac{1}{2} & -\frac{1}{8} \\ 0 & 1 & 0 & \vdots & -\frac{15}{8} & \frac{1}{2} & \frac{3}{8} \\ 0 & 0 & 1 & \vdots & \frac{5}{4} & 0 & -\frac{1}{4} \end{vmatrix}$

Step 3. Since $C=I_3$, then $D=A^{-1}$ $\qquad A^{-1} = \begin{matrix} \frac{13}{8} & -\frac{1}{2} & -\frac{1}{8} \\ -\frac{15}{8} & \frac{1}{2} & \frac{3}{8} \\ \frac{5}{4} & 0 & -\frac{1}{4} \end{matrix}$

Therefore, Matrix A is nonsingular.

Theorem 2.10. An *nxn* matrix is nonsingular if and only if it is row equivalent to I_n

Example 4.

$$\text{Let } A = \begin{vmatrix} 1 & 2 & 3 \\ 1 & 2 & 1 \\ 5 & 2 & 3 \end{vmatrix}$$

$$\text{Step 1.} \begin{vmatrix} 1 & 2 & -3 & \vdots & 1 & 0 & 0 \\ 1 & -2 & 1 & \vdots & 0 & 1 & 0 \\ 5 & -2 & -3 & \vdots & 0 & 0 & 1 \end{vmatrix}$$

Step. Transform to reduced row echelon form.

$$\begin{vmatrix} 1 & 2 & -3 & \vdots & 1 & 0 & 0 \\ 0 & 1 & -1 & \vdots & \frac{1}{4} & -\frac{1}{4} & 0 \\ 0 & 0 & 0 & \vdots & -2 & -3 & 1 \end{vmatrix}$$

Step 3. Since C has a row of zeros, then A is a singular matrix and A^{-1} does not exist.

Note that at this point, we have shown that the following statements are equivalent for an *n x n MATRIX* a:

1. A is nonsingular.
2. Ax = 0 has only the trivial solution.
3. A is row (column) equivalent to I_n (the reduced row echelon form of A is I_n)
4. The linear system Ax=b has a unique solution for every n X 1 matrix b.
5. A is a product of elementary matrices.

Lesson 2.4. Equivalent Matrices

DEFINITION 2.5

If A and B are two *m x n* matrices, then A is equivalent to *B* if we obtain B from A by afinite sequence of elementary row or elementary column operations.

Theorem 2.12. if A is any nonzero *m x n* matrix, then A is equivalent to a partitionedmatrix of the form.

Example 1. Let $A = \begin{vmatrix} 1 & 1 & 2 & -1 \\ 1 & 2 & 1 & 0 \\ -1 & -4 & 1 & -2 \\ 1 & -2 & 5 & -4 \end{vmatrix}$

To find a matrix of the form described in theorem 2.12, which is equivalent to A, weproceed as follows. Apply row operation -1r_1 + r_2 $\rightarrow r_2$ to obtain

$$\begin{vmatrix} 1 & 1 & 2 & -1 \\ 0 & 1 & -1 & 1 \\ -1 & -4 & 1 & -2 \\ 1 & -2 & 5 & -4 \end{vmatrix}$$ Apply $1r_1 + r_3 \rightarrow r_3$

$$\begin{vmatrix} 1 & 1 & 2 & -1 \\ 0 & 1 & -1 & 1 \\ 0 & -3 & 3 & -3 \\ 1 & -2 & 5 & -4 \end{vmatrix}$$ Apply -$1r_1 + r_4 \rightarrow r_4$

$$\begin{vmatrix} 1 & 1 & 2 & -1 \\ 0 & 1 & -1 & 1 \\ 0 & -3 & 3 & -3 \\ 0 & -3 & 3 & -3 \end{vmatrix}$$ Apply $-1r_3 + r_4 \rightarrow r_4$

$$\begin{vmatrix} 1 & 1 & 2 & -1 \\ 0 & 1 & -1 & 1 \\ 0 & -3 & 3 & -3 \\ 0 & 0 & 0 & 0 \end{vmatrix}$$ Apply $-\frac{1}{3}r_3 \rightarrow r_3$

$$\begin{vmatrix} 1 & 1 & 2 & -1 \\ 0 & 1 & -1 & 1 \\ 0 & 1 & -1 & 1 \\ 0 & 0 & 0 & 0 \end{vmatrix}$$

Apply -1$r_2 + r_3 \rightarrow r_3$

$$\begin{vmatrix} 1 & 1 & 2 & -1 \\ 0 & 1 & -1 & 1 \\ 0 & 0 & 0 & 0 \\ 0 & 0 & 0 & 0 \end{vmatrix}$$

Apply $-1r_2 + r_1 \rightarrow r_1$

$$\begin{vmatrix} 1 & 0 & 3 & -2 \\ 0 & 1 & -1 & 1 \\ 0 & 0 & 0 & 0 \\ 0 & 0 & 0 & 0 \end{vmatrix}$$

Apply $-3c_1 + c_3 \rightarrow c_3$

$$\begin{vmatrix} 1 & 0 & 0 & -2 \\ 0 & 1 & -1 & 1 \\ 0 & 0 & 0 & 0 \\ 0 & 0 & 0 & 0 \end{vmatrix}$$

Apply $2c_1 + c_4 \rightarrow c_4$

$$\begin{vmatrix} 1 & 0 & 0 & 0 \\ 0 & 1 & -1 & 1 \\ 0 & 0 & 0 & 0 \\ 0 & 0 & 0 & 0 \end{vmatrix}$$

Apply $1c_2 + c_3 \rightarrow c_4$

$$\begin{vmatrix} 1 & 0 & 0 & 0 \\ 0 & 1 & 0 & 0 \\ 0 & 0 & 0 & 0 \\ 0 & 0 & 0 & 0 \end{vmatrix}$$

This is the desired matrix.

EXERCISES 2.1

A.) Solve the system in each given linear system is in row echelon form.

1.)
$$x + 3y + z = = 6$$
$$y + z = 5$$
$$z = 4$$

2.)
$$x - 4y + 5z + w = 0$$
$$z - 2w = 4$$
$$w = 1$$

3.)
$$x + 2y - z + 3w = 4$$
$$w = 5$$

4.)
$$5x - 6y - 7x = 7$$
$$6x - 4y + 10z = -34$$
$$2x + 4y - 3z = 29$$

B.) Find all solutions, if any exist, by using Gauss-Jordan reduction method of the followinglinear system:

5.)
$$x + 2y + 3 = -1$$
$$x - 2y - z = -1$$
$$3x + y + z = 3$$

6.)
$$x + y + 3z + 2w = 13$$
$$x - 2y + 2z + w = 8$$
$$3x + y + z - 2w = 1$$

7.)
$$x + y - 2z = 1$$
$$x - 2y - 3z = 4$$
$$2w + 4y - 3z = 29$$

C.) Which of the given matrices are singular? For the non-singular ones, find theinverse.

8. $\begin{bmatrix} 2 & 4 \\ 1 & 2 \end{bmatrix}$

9. $\begin{bmatrix} 2 & 4 \\ 1 & -2 \end{bmatrix}$

10. $\begin{bmatrix} 1 & 2 & 3 \\ 1 & 1 & 2 \\ 0 & 1 & 2 \end{bmatrix}$

ASSESSMENT

A. Find the inverse of the following:

1. $A = \begin{bmatrix} 1 & 4 \\ 2 & 3 \end{bmatrix}$

2. $A = \begin{bmatrix} 1 & 0 & 2 \\ 0 & 1 & -1 \end{bmatrix}$

B. Find the inverse, if it exists, of the following

2. $A = \begin{bmatrix} 1 & 1 & 2 & 1 \\ 0 & -2 & 0 & 0 \\ 1 & 2 & 1 & -2 \\ 0 & 3 & 2 & 1 \end{bmatrix}$

4. $A = \begin{bmatrix} 1 & 1 & 5 & 1 \\ 2 & 3 & 9 & 1 \\ -1 & 1 & 1 & 1 \\ 1 & 2 & 6 & 1 \end{bmatrix}$

C. Which of the following homogeneous systems have a nontrivial solution?

6. $\begin{aligned} 2x \quad\quad + z &= 0 \\ 2x + 2y - 4z &= 0 \\ 4x + 2y + 3z &= 0 \end{aligned}$

6. $\begin{aligned} 2x + y - z &= 0 \\ x - 2y - 3z &= 0 \\ -3x - y + 2z &= 0 \end{aligned}$

7. $\begin{aligned} 3x + y + 3z &= 0 \\ -2x + 2y - 4z &= 0 \\ 2x - 3y + 5z &= 0 \end{aligned}$

8. $\begin{aligned} 3x + y + 2z &= 0 \\ -x + y + z &= 0 \\ x + 2y + z &= 0 \end{aligned}$

8. $\begin{aligned} x - y + z &= 0 \\ 2x + y \quad\quad &= 0 \\ 2x - 2y + 2z &= 0 \end{aligned}$

10. $\begin{aligned} 2x - y + 5z &= 0 \\ 3x + 2y - 3z &= 0 \\ x - y + 4z &= 0 \end{aligned}$

KEY TO CORRECTION

Exercises

1. $x = -1$
   $y = 1$
   $z = 4$

2. $x = 4y - 31$
   $z = 6$
   $w = 1$

3. $x + 2y - z = -11$
   $w = 5$

4. $x = 2$
   $y = -1$
   $z = 0$

5. $x = 2$
   $y = 2$
   $z = -2$

6. $X - w = -\frac{14}{11}, y = \frac{1}{11}, z + w = \frac{52}{11}$

7. $x = 2, y = -1, z = 0$

8. Singular Matrix

9. $\begin{bmatrix} \frac{1}{4} & \frac{1}{2} \\ \frac{1}{8} & -\frac{1}{4} \end{bmatrix}$

10. $\begin{bmatrix} 0 & 1 & -1 \\ 2 & -2 & -1 \\ -1 & 1 & 1 \end{bmatrix}$

Assessment

1. $\begin{bmatrix} -\frac{3}{5} & \frac{4}{5} \\ \frac{2}{5} & -\frac{1}{5} \end{bmatrix}$

2. $\begin{bmatrix} 1 & -1 & 0 & -1 \\ 0 & -1 & 0 & 0 \\ -\frac{1}{5} & & \frac{1}{5} & \frac{3}{5} \\ -\frac{2}{5} & -\frac{1}{2} & -\frac{2}{5} & -\frac{1}{5} \end{bmatrix}$

3. $x - \frac{1}{2}z = 0, y + \frac{1}{2}z = 0$
   Nontrivial Solution

6. $x = 0, y = 0, z = 0$
Trivial Solution

9. $x + \frac{1}{3}z = 0, y - \frac{2}{3}z = 0$
   Nontrivial Solution

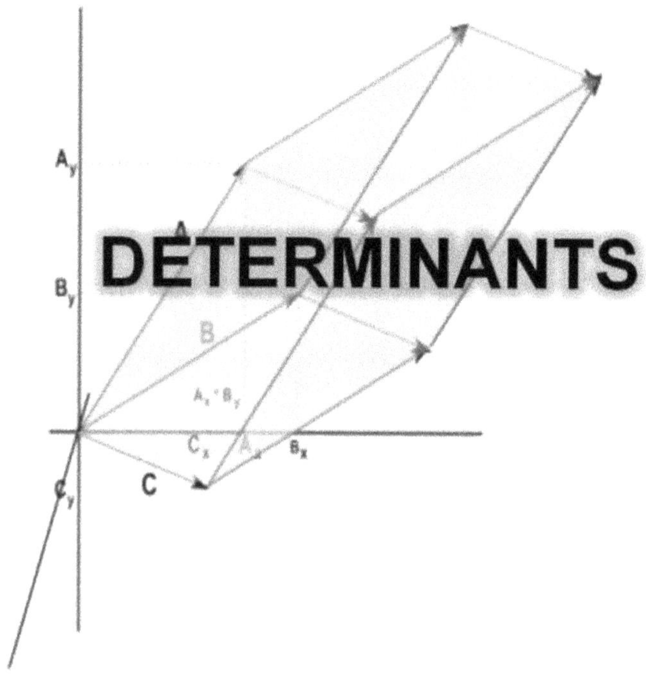

source: euclideanSpace

DETERMINANTS

OBJECTIVES

After working through this section, you should be able to:

- Understand and solve linear systems by Cramer's rule.
- Understand and define determinants
- Simplify the properties of determinants
- Solve determinants via reduction to triangular form
- Evaluate determinants.
- Find the inverse matrix.
- Apply Cramer's rule for solving a system of linear equations, if the determinant ofthe matrix of coefficients of the system is not zero[1]

Lesson 3.1. Definition

INTRODUCTION

Another very important number associated with a square matrix A is the determinant of A, which we now define. Determinants first arose in the solution of linear systems. Although the methods given in chapter 2 for solving such systems are more efficient thanthose involving determinants, determinants will be useful for our further study of a linear transformation $L:V \rightarrow V$ in Chapter 6. First, we deal briefly with permutations used in our definition of determinant. When we use the term matrix throughout this chapter, we mean square matrix.

DISCUSSION

DEFINITION 3.1

Let S= { 1,2,...,n} be the set of integers from 1 to n, arranged in ascending order. A rearrangement j1j2 ...jn of the elements of S is called a **permutation** of S. We can considera permutation of S to be a one-to-one mapping of S onto itself. Thus, 4321 is a permutation of S={1,2,3,4}. [2]

n(n-1)(n-2).....2(1) denoted by n! (n factorial)

Inversion- if a larger integer precedes a smaller integer in a permutation.

Even and Odd permutation- According to whether the total number of inversions in it is odd or even.

n!	factor	Results
1!	1x1	1
2!	2x1	2
3!	3x2x1	6
4!	4x3x2x1	24
5!	5x4x3x2x1	120
6!	6x5x4x3x2x1	720
7!	7x6x5x4x3x2x1	5,040
8!	8x7x6x5x4x3x2x1	40,320
9!	9x8x7x6x5x4x3x2x1	362,880
10!	10x9x8x7x6x5x4x3x2x1	3,628,800

Example 1.

S1= {1} n!= 1! = 1 permutation S2= {1,2} 2!= 2 { 12,21}

S3= {1,2,3}= 3!= 6 { 123,321,231,312,213,132}

S1 , 1 (0 inversion)- Even permutation

S2 , 12 (0 inversion) – Even permutation
21 (1 inversion) -odd permutation

S3 , 123 0 inversion) 231 (2 inversions) 312 (2 inversions) = even permutations

132 (1 inversion) 213(1 inversion) 321 (3 inversions) = Odd permutations

DEFINITION 3.2

Let A = [Aij] be an *nxn* matrix. we define the determinant of A by $|A| = \sum(\pm)A_{ij}A_{2j2}\cdots A_{njn}$, where the summation ranges over all permutations $j_1j_2\cdots\cdots j_n$ of the set S={1,2,...n}. The sign is taken as + or – according to whether the permutation is even or odd.

Example 2.

If $A = [a_{11}]$ is a 1 x 1 matrix, then $|A| = a_{11}$

Example 3.

If $A = \begin{bmatrix} a_{11} & a_{12} \\ a_{21} & a_{22} \end{bmatrix}$, then to obtain ä$A$ä, we write down the terms $a_1 - a_2$ and replace the dashes with all possible elements S2: The subscripts become 12 and 21. Now 12 is an even permutation and 21 is an odd permutation. Thus $|A| = a_{11}a_{22} - a_{12}a_{21}$. (Read as a sub 1 and 1, a sub 2 and 2 ……a sub n and n). 3

Hence, we see that $[A]$ can be obtained by forming the product of the entries on the line from left to right and subtracting from this number the product of the entries on the line from right to left.[3]

$$\begin{matrix} a_{11} & a_{12} \\ a_{21} & a_{22} \end{matrix}$$

Thus, if $A = \begin{bmatrix} 2 & -3 \\ 4 & 5 \end{bmatrix}$, then $A = (2)(5) - (-3)(4) = 22$

Example 4. If $A = \begin{bmatrix} a_{11} & a_{12} & a_{13} \\ a_{21} & a_{22} & a_{23} \\ a_{31} & a_{32} & a_{33} \end{bmatrix}$

To compute $|A|$ we write down the six terms

S_3,

Even (+)	Odd (-)
a_{11}, a_{22}, a_{33}	a_{11}, a_{23}, a_{32}
a_{12}, a_{23}, a_{31}	a_{12}, a_{21}, a_{33}
a_{13}, a_{21}, a_{32}	a_{13}, a_{22}, a_{31}

$$|A| = a_{11}a_{22}a_{33} + a_{12}a_{23}a_{31} + a_{13}a_{21}a_{32} - a_{11}a_{23}a_{32} - a_{12}a_{21}a_{33} - a_{13}a_{22}a_{31}$$

We can also obtain $|A|$ as follows. Repeat the first and second columns of A, shown next. Form the sum of the products of the entries on the lines from left to right, and subtract from this number the products of the entries on the lines from right to left (verify).

Basket Method

$$\begin{matrix} a_{11} & a_{12} & a_{13} & a_{11} & a_{12} \\ a_{21} & a_{22} & a_{23} & a_{21} & a_{22} \\ a_{31} & a_{32} & a_{33} & a_{31} & a_{32} \end{matrix}$$

$$|A| = a_{11}a_{22}a_{33} + a_{12}a_{23}a_{31} + a_{13}a_{21}a_{32} - a_{13}a_{22}a_{31} - a_{11}a_{23}a_{32} - a_{12}a_{21}a_{33}$$

Example 5.

Let $A = \begin{vmatrix} 1 & 2 & 3 \\ 2 & 1 & 3 \\ 3 & 1 & 2 \end{vmatrix}$ Evaluate $|A|$

Solution: Substituting (1); we find that

$$\begin{vmatrix} 1 & 2 & 3 \\ 2 & 1 & 3 \\ 3 & 1 & 2 \end{vmatrix} = (1)(1)(2) + (2)(3)(3) + (3)(2)(1) - (1)(3)(1) - (2)(2)(2) - (3)(1)(3) = 2 + 18 + 6 - 3 - 8 - 9 = 6$$

We obtain the same result by using the easy method (Basket Method).

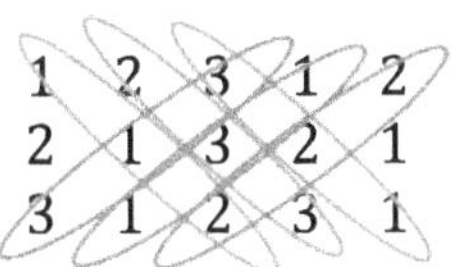

$$|A| = (1)(1)(2) + (2)(3)(3) + (3)(2)(1) - (1)(3)(1) - (2)(2)(2) - (3)(1)(3)$$

$$= 2 + 18 + 6 - 3 - 8 - 9 = 6$$

Lesson 3.2. Properties of Determinants

Theorem 3.1 The determinants of a matrix and its transpose are equal. [4]

Example 1. $A = \begin{vmatrix} 1 & 2 & 3 \\ 2 & 1 & 3 \\ 3 & 1 & 2 \end{vmatrix}$ $\quad |A| = 6 \qquad A^T = \begin{vmatrix} 1 & 2 & 3 \\ 2 & 1 & 1 \\ 3 & 3 & 2 \end{vmatrix}$ $\quad |A^T| = 6$

Theorem 3.2. If matrix B results from matrix A by interchanging two different rows (columns) of A, then $|B| = |A|$[2]

Example 2. We have $|A| = \begin{vmatrix} 2 & -1 \\ 3 & 2 \end{vmatrix} = -\begin{bmatrix} 3 & 2 \\ 2 & -1 \end{bmatrix} = \begin{bmatrix} 2 & 3 \\ -1 & 2 \end{bmatrix} = |A^T| = 7$

Theorem 3.3 If two rows (columns) of A are equal, then $|A| = 0$. [2]

Example 3. We have $\begin{vmatrix} 1 & 2 & 3 \\ -1 & 0 & 7 \\ 1 & 2 & 3 \end{vmatrix} = 0$

Solution: $|A| = (0) + (14) + (-6) - (0) - (14) - (-6) = 8 - 8 = 0$

Theorem 3.4. If a row (column) of A consists entirely of zeros, then $|A| = 0$. [5]

Example 4. We have $\begin{vmatrix} 1 & 2 & 3 \\ 4 & 5 & 6 \\ 0 & 0 & 0 \end{vmatrix}$ $|A| = 0 \qquad \begin{vmatrix} 0 & 1 & 4 \\ 0 & 2 & 5 \\ 0 & 3 & 6 \end{vmatrix}$ $|A| = 0$

If B is obtained from A by multiplying a row (column) of A by a
Theorem 3.5 real number k, then $|B| = k|A|$. [2]

Example 5. We have

$$\begin{vmatrix} 2 & 6 \\ 1 & 12 \end{vmatrix} = 2\begin{vmatrix} 1 & 3 \\ 1 & 12 \end{vmatrix} = (2)(3)\begin{vmatrix} 1 & 1 \\ 1 & 4 \end{vmatrix} = 6(4 - 1) = 18$$

We can use Theorem 3.5 to simplify the computation of $|A|$ by factoring out common factors from rows and columns of A.

Example 6. We have $\begin{vmatrix} 1 & 2 & 3 \\ 1 & 5 & 3 \\ 2 & 8 & 6 \end{vmatrix} = 2\begin{vmatrix} 1 & 2 & 3 \\ 1 & 5 & 3 \\ 1 & 4 & 3 \end{vmatrix} = (2)(3)\begin{vmatrix} 1 & 2 & 1 \\ 1 & 5 & 1 \\ 1 & 4 & 1 \end{vmatrix}$

$$= (2)(3)(0) = 0$$

Here, we first factored out 2 from the third row and 3 from the third column, and then used Theorem 3.3., since the first and third columns are equal.

Theorem 3.6. If $B = [b_{ij}]$ is obtained from $A = [a_{ij}]$ by adding to each element of the rth row (column) of A,k times the corresponding element of the sth row (column), $r \neq s$ of A, then $|B| = |A|$. [4]

Example 7.

$$A = \begin{vmatrix} 1 & 2 & 3 \\ 2 & -1 & 3 \\ 1 & 0 & 1 \end{vmatrix} \quad |A| = 4 \quad B = \begin{vmatrix} 5 & 0 & 9 \\ 2 & -1 & 3 \\ 1 & 0 & 1 \end{vmatrix} \quad |B| = 4$$

$\begin{vmatrix} 1 & 2 & 3 \\ 2 & -1 & 3 \\ 1 & 0 & 1 \end{vmatrix} = \begin{vmatrix} 5 & 0 & 9 \\ 2 & -1 & 3 \\ 1 & 0 & 1 \end{vmatrix}$ obtained by adding twice the second row to the first row. By applying the definition of determination to the first and second determinants, both are seen to have the value 4.

Theorem 3.7. If a matrix $A = [a_{ij}]$is upper (lower) triangular, then $|A| = a_{11}a_{22}...a_{nn}$ that is, the determinant of the triangular matrix is the product of the elements on themain diagonal.

Example 8. Let $A = \begin{vmatrix} 4 & 3 & 2 \\ 3 & -2 & 5 \\ 2 & 4 & 6 \end{vmatrix}$.Compute $|A|$.

Solution

We have $|A| = 2\ (A\frac{1}{2}rr_3 \rightarrow r_3)$

$$= 2\left(\begin{vmatrix} 4 & 3 & 2 \\ 3 & -2 & 5 \\ 1 & 2 & 3 \end{vmatrix}\right) \qquad \text{Multiply 3 by } \frac{1}{2}$$

We have

$$= (-1)2\left(\begin{vmatrix} 4 & 3 & 2 \\ 3 & -2 & 5 \\ 1 & 2 & 3 \end{vmatrix}\right) r_1 \leftrightarrow r_3 \quad \text{Interchange rows 1 and 3}$$

We have

$$= -2\left(\begin{vmatrix} 1 & 2 & 3 \\ 3 & -2 & 5 \\ 4 & 3 & 2 \end{vmatrix}\right) \quad -3r_1 + r_2 \to r_2 \text{and} -4r_1 + r_3 \to r_3$$ Zero out below the (1,1) entry

We have

$$= -2\left(\begin{vmatrix} 1 & 2 & 3 \\ 0 & -8 & -4 \\ 0 & -5 & -10 \end{vmatrix}\right)$$

$$= -2\left(\begin{vmatrix} 1 & 2 & 3 \\ 0 & -8 & -4 \\ 0 & -5 & -10 \end{vmatrix}\right) \quad -\tfrac{5}{8}r_2 + r_3 \to r_3$$ Zero out below the (2,2) entry

$$= -2\left(\begin{vmatrix} 1 & 2 & 3 \\ 0 & -8 & -4 \\ 0 & 0 & -\frac{30}{4} \end{vmatrix}\right)$$

Next, we compute the determinant of the upper triangular matrix. Using the reduction totriangular form.[6]

$|A| = -2(1)(-8)(-\tfrac{30}{4}) = -120$

Next, we prove that the determinant of a product of two matrices is the product of their determinants and that A is nonsingular if and only if $|A| \neq 0$.

Lemma 3.1. If E is an elementary matrix, then $|EA| = (|E|)(|A|)$ and $(|A|)(|E|)$.

Theorem 3.8. If A is an n x n matrix, then A is nonsingular if and only if $|A| \neq 0$. [6]

Corollary 3.1. If A is an n x n matrix, then Ax=0 has a nontrivial solution if and only if $|A| = 0$.

Example 9. Let A be a 4 x 4 matrix with $|A| = -2$.

a.) Describe the set of all solutions to the homogenous system Ax=0.
b.) If A is transformed to a reduced row echelon form B, what is B?
c.) Give an expression for a solution to the linear system Ax=b,

where $b = 0\begin{bmatrix}1\\2\\3\\4\end{bmatrix}$

d.) Can the linear system Ax= b has more than one solution? Explain.

e.) Does A^{-1} exist?

Solution

a. Since $|A| \neq 0$, by Corollary 3.1, the homogenous system has only the trivial solution.

b. Since $|A| \neq 0$, by Corollary 2.2 in Section 2.3, A is a nonsingular matrix, and by Theorem 2.2, $B = l_n$

c. A solution to the given system is given by x= $A^{-1}b$.

d. No. The solution given in part © is the only one

e. Yes.

Theorem 3.9. If A and B are n x n matrices, then $|AB| = |A| + |B|$

Example 10. Let $A = \begin{vmatrix}1 & 2\\3 & 4\end{vmatrix}$ and $B = \begin{vmatrix}2 & -1\\1 & 2\end{vmatrix}$

Then $|A| = -2$ and $|B| = 5$

On the other hand, $AB = \begin{vmatrix}4 & 3\\10 & 5\end{vmatrix}$ and $|AB| = -10 = |A||B|$

Corollary 3.2. If A is nonsingular, then $|A^{-1}| = \frac{1}{|A|}$

Corollary 3.3. If A and B are similar matrices, then $|A| = |B|$

Example 11. Let $A \begin{vmatrix} & 2 & 2 & 3\\ = & 0 & 3 & 4\\ & 0 & 2 & 4\end{vmatrix}$ $B = \begin{vmatrix}2 & 2 & 3\\0 & 3 & 4\\1 & -2 & -4\end{vmatrix}$ $C = \begin{vmatrix}2 & 2 & 3\\0 & 3 & 4\\1 & 0 & 0\end{vmatrix}$

Then $|A| = 8, |B| = -9\ and\ |C| = -1, so\ |C| = |A| + |B|$

Lesson 3.3. Cofactor Expansion

Definition 3.3. Let $A = [a_{ij}]$ be an n x n matrix. Let M_{ij} be the (n-1) x (n-1) submatrix of A obtained by deleting the ith row and jth column of A. The determinant M_{ij} is called the motor of a_{ij}.[5]

Definition 3.4. Let $A = |a_{ij}|$ be an n x n matrix. The cofactor A_{ij} of a_{ij} is defined as $A_{ij} = (-1)^{i+j}|M_{ij}|$.

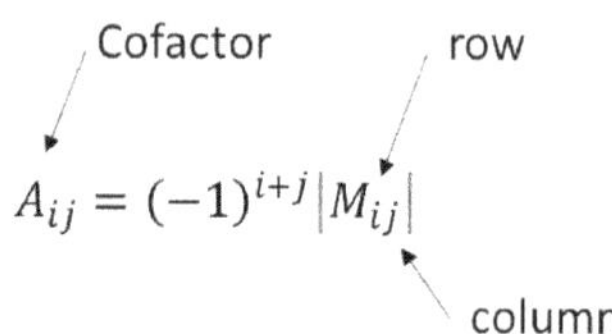

$$A_{ij} = (-1)^{i+j}|M_{ij}|$$

Example 1. Let $A = \begin{vmatrix} 3 & -1 & 2 \\ 4 & 5 & 6 \\ 7 & 1 & 2 \end{vmatrix}$ $\quad nM_{12} = \begin{vmatrix} 4 & 6 \\ 7 & 2 \end{vmatrix} = 8 - 42 = -34$

$\begin{vmatrix} 3 & -1 & 2 \\ 4 & 5 & 6 \\ 7 & 1 & 2 \end{vmatrix}$ $\quad M_{23} = \begin{vmatrix} 3 & -1 \\ 7 & 1 \end{vmatrix} = 3 - (-7) = 10$

$\begin{vmatrix} 3 & -1 & 2 \\ 4 & 5 & 6 \\ 7 & 1 & 2 \end{vmatrix}$ $\quad M_{31} = \begin{vmatrix} -1 & 2 \\ 5 & 6 \end{vmatrix} = -6 - 10 = -16. Also$

$$A_{ij} = (-1)^{i+j}M_{ij}$$

$A_{12} = (-1)^{1+2}M_{ij} = (-1)(-34) = 34$

$A_{23} = (-1)^{2+3}|M_{23}| = (-1)(10) = -10$

And $A_{31} = (-1)^{3+1}|M_{31}| = (1)(-16)$

Theorem 3.10. Let $A = |a_{ij}|$ be an **n x n matrix**. Then

$|A| = a_{i1}A_{i1} + a_{i2}A_{i2} + \cdots + a_{in}A_{in}$ $[Expansion\ of\ |A|\ along\ the\ ith\ row]$

And

$|A| = a_{ij}A_{ij} + a_{2j}A_{2j} + \cdots + a_{nj}A_{nj}$$[Expansion\ of\ |A|\ along\ the\ jth\ column]$

Example 2. To evaluate the determinant $\begin{vmatrix} 1 & 2 & -3 & 4 \\ -4 & 2 & 1 & 3 \\ 3 & 0 & 0 & -3 \\ 2 & 0 & -2 & 3 \end{vmatrix}$

Solution: Expanding the third row.

$$\begin{vmatrix} 1 & 2 & -3 & 4 \\ 4 & 2 & 1 & 3 \\ 3 & 0 & 0 & -3 \\ 2 & 0 & -2 & 3 \end{vmatrix} = (-1)^{3+1}(3)\begin{vmatrix} 2 & -3 & 4 \\ 2 & 1 & 3 \\ 0 & -2 & 3 \end{vmatrix} +$$

$$\begin{vmatrix} 1 & 2 & -3 & 4 \\ -4 & 2 & 1 & 3 \\ 3 & 0 & 0 & -3 \\ 2 & 0 & -2 & 3 \end{vmatrix} = (-1)^{3+2}(0)\begin{vmatrix} 1 & -3 & 4 \\ 4 & 1 & 3 \\ 2 & -2 & 3 \end{vmatrix} +$$

$$\begin{vmatrix} 1 & 2 & -3 & 4 \\ 4 & 2 & 1 & 3 \\ 3 & 0 & 0 & -3 \\ 2 & 0 & -2 & 3 \end{vmatrix} = (-1)^{3+3}(0)\begin{vmatrix} 1 & 2 & 4 \\ -4 & 2 & 3 \\ 2 & 0 & 3 \end{vmatrix} +$$

$$\begin{vmatrix} 1 & 2 & -3 & 4 \\ 4 & 2 & 1 & 3 \\ 3 & 0 & 0 & -3 \\ 2 & 0 & -2 & 3 \end{vmatrix} = (-1)^{3+4}(-3)\begin{vmatrix} 1 & 2 & -3 \\ -4 & 2 & 1 \\ 2 & 0 & -2 \end{vmatrix}$$

$$= (+1)(3)(20) + 0 + 0 + (-1)(-3)(-4) = \mathbf{48}$$

Lesson 3.4 Inverse of a Matrix

We saw in Section 3.3. that theorem 3.10 provides formulas for expanding $|A|$ along either a row or a column of A, we obtain another method for finding the inverse of anonsingular matrix.

Theorem 3.11. If $A = [a_{ij}]$is an n X n matrix, then

$$a_{i1}A_{k1} + a_{i2}A_{k2} + \cdots + a_{in}A_{kn} = 0\ for\ i \neq k;$$

$$a_{1j}A_{1k} + a_{2j}A_{2k} + \cdots + a_{nj}A_{nk} = 0\ for\ j \neq k.$$

$$0 = |A| = a_{i1}A_{k1} + a_{i2}A_{k2} + \cdots + a_{in}A_{kn}$$

Example 1. Let $A = \begin{vmatrix} 1 & 2 & 3 \\ -2 & 3 & 1 \\ 4 & 5 & -2 \end{vmatrix}$ $Then\ A_{21} = (-1)^{2+1}\begin{vmatrix} 2 & 3 \\ 5 & -2 \end{vmatrix} = 19$

$$A_{22} = (-1)^{2+2}\begin{vmatrix} 1 & 3 \\ 4 & -2 \end{vmatrix} = -14, and\ A_{23} = (-1)^{2+3}\begin{vmatrix} 1 & 2 \\ 4 & 5 \end{vmatrix} = 3. Now$$

$$a_{31}A_{21} + a_{32}A_{22} + a_{33}A_{23} = (4)(19) + (5)(-14) + (-2)(3) = 0, and$$

$$a_{11}A_{21} + a_{12}A_{22} + a_{13}A_{23} = (1)(19) + (2)(-14) + (3)(3) = 0$$

Definition 3.5. Let $A = If\ A = [a_{ij}]$ is an n x n matrix, then n x n matrix adj A, called the adjoint of A, is the matrix whose (I,j)th entry is the cofactorA_{ji}, of a_{ji}.[5] Thus

$$adj\ A = \begin{vmatrix} A_{11} & A_{21} & \cdots & A_{n1} \\ A_{21} & A_{22} & \cdots & A_{n2} \\ \vdots & \vdots & \cdots & \vdots \\ A_{1n} & A_{2n} & \cdots & A_{nn} \end{vmatrix}$$

Example 2. Let $A = \begin{vmatrix} 3 & -2 & 1 \\ 5 & 6 & 2 \\ 1 & 0 & -3 \end{vmatrix}$ Compute adj A

Solution

$\begin{vmatrix} 3 & -2 & 1 \\ 5 & 6 & 2 \\ 1 & 0 & -3 \end{vmatrix}$ $A_{11} = (-1)^{1+1}\begin{vmatrix} 6 & 2 \\ 0 & -3 \end{vmatrix} = -18$

$$\begin{vmatrix} 3 & -2 & 1 \\ 5 & 6 & 2 \\ 1 & 0 & -3 \end{vmatrix} \quad A_{12} = (-1)^{1+2} \begin{vmatrix} 5 & 2 \\ 1 & -3 \end{vmatrix} = 17$$

$$\begin{vmatrix} 3 & -2 & 1 \\ 5 & 6 & 2 \\ 1 & 0 & -3 \end{vmatrix} \quad A_{13} = (-1)^{1+3} \begin{vmatrix} 5 & 6 \\ 1 & 0 \end{vmatrix} = -6$$

$$\begin{vmatrix} 3 & -2 & 1 \\ 5 & 6 & 2 \\ 1 & 0 & -3 \end{vmatrix} \quad A_{21} = (-1)^{2+1} \begin{vmatrix} -2 & 1 \\ 0 & -3 \end{vmatrix} = -6$$

$$\begin{vmatrix} 3 & -2 & 1 \\ 5 & 6 & 2 \\ 1 & 0 & -3 \end{vmatrix} \quad A_{22} = (-1)^{2+2} \begin{vmatrix} 3 & 1 \\ 1 & -3 \end{vmatrix} = -10$$

$$\begin{vmatrix} 3 & -2 & 1 \\ 5 & 6 & 2 \\ 1 & 0 & -3 \end{vmatrix} \quad A_{23} = (-1)^{2+3} \begin{vmatrix} 3 & -2 \\ 1 & 0 \end{vmatrix} = -2$$

$$\begin{vmatrix} 3 & -2 & 1 \\ 5 & 6 & 2 \\ 1 & 0 & -3 \end{vmatrix} \quad A_{31} = (-1)^{3+1} \begin{vmatrix} -2 & 1 \\ 6 & 2 \end{vmatrix} = -10$$

$$\begin{vmatrix} 3 & -2 & 1 \\ 5 & 6 & 2 \\ 1 & 0 & -3 \end{vmatrix} \quad A_{32} = (-1)^{3+2} \begin{vmatrix} 3 & 1 \\ 5 & 2 \end{vmatrix} = -1$$

$$\begin{vmatrix} 3 & -2 & 1 \\ 5 & 6 & 2 \\ 1 & 0 & -3 \end{vmatrix} \quad A_{33} = (-1)^{3+3} \begin{vmatrix} 3 & -2 \\ 5 & 6 \end{vmatrix} = 28$$

$$adj\ A = \begin{vmatrix} A_{11} & A_{21} & \cdots & A_{n1} \\ A_{12} & A_{22} & \cdots & A_{n2} \\ \vdots & \vdots & \cdots & \vdots \\ A_{1n} & A_{2n} & \cdots & A_{nn} \end{vmatrix} \qquad adj\ A = \begin{vmatrix} -18 & -6 & -10 \\ 17 & -10 & -1 \\ -6 & -2 & 28 \end{vmatrix}$$

Theorem 3.12. If $A = [a_{ij}]$ is an n x n matrix, then $A\,(adj\ A) = (adj\ A)A = |A|l_{n.}$

Proof

We have

$$A(adj\ A) = \begin{bmatrix} a_{11} & a_{12} & \cdots & a_{1n} \\ a_{21} & a_{22} & \cdots & a_{2n} \\ \vdots & \vdots & \cdots & \vdots \\ a_{i1} & a_{i2} & \cdots & a_{in} \\ \vdots & \vdots & & \vdots \\ a_{n1} & a_{n2} & \cdots & a_{nn} \end{bmatrix} \begin{bmatrix} A_{11} & A_{12} & \cdots & A_{j1} & \cdots & A_{n1} \\ A_{21} & A_{22} & \cdots & A_{j2} & \cdots & A_{n2} \\ \vdots & \vdots & & \vdots & & \vdots \\ A_{1n} & A_{2n} & \cdots & A_{jn} & \cdots & A_{nn} \end{bmatrix}$$

The *(i,j)th* element in the product matrix A (adj A) is, by Theorem 3.10,
$a_{i1}A_{j1} + a_{i2}A_{j2} + \cdots + a_{in}A_{jn} = |A|$ if $i = j$

=0 if i≠j

This means that A (adj A) = $\begin{vmatrix} |A| & 0 & \cdots & 0 \\ 0 & |A| & & 0 \\ \vdots & \vdots & \vdots & \vdots \\ 0 & \cdots & 0 & |A| \end{vmatrix} = |A|$

The (i.j)th element in the product matrix (adj A)A is, by Theorem 3.10,

$$A_{2i}a_{1j} + A_{2i}a_{2j} + \cdots + A_{ni}a_{nj} = |A| \quad if\ i = j$$

$$= 0 \quad if\ i \neq j$$

Thus $(adj\ A)A = |A|$

Example 3. Consider the matrix of Example 2. Then

$$\begin{vmatrix} 3 & -2 & 1 \\ 5 & 6 & 2 \\ 1 & 0 & -3 \end{vmatrix} \begin{vmatrix} -18 & -6 & 10 \\ 17 & -10 & -1 \\ -6 & -2 & 28 \end{vmatrix} = \begin{vmatrix} -94 & 0 & 0 \\ 0 & -94 & 0 \\ 0 & 0 & -94 \end{vmatrix} = -94 \begin{vmatrix} 1 & 0 & 0 \\ 0 & 1 & 0 \\ 0 & 0 & 1 \end{vmatrix}$$

and $\begin{vmatrix} -18 & -6 & -10 \\ 17 & -10 & -1 \\ -6 & -2 & 28 \end{vmatrix} \begin{vmatrix} 3 & -2 & 1 \\ 5 & 6 & 2 \\ 1 & 0 & -3 \end{vmatrix} = -94 \begin{vmatrix} 1 & 0 & 0 \\ 0 & 1 & 0 \\ 0 & 0 & 1 \end{vmatrix}$

We now have a new method for finding the inverse of a nonsingular matrix, and we state this result as the following corollary.

Corollary 3.4 If A is an n x n matrix and $|A| \neq 0, then$

$$A^{-1} = \frac{1}{|A|}(adj\ A) = \begin{bmatrix} \frac{A_{11}}{|A|} & \frac{A_{21}}{|A|} & \cdots & \frac{A_{n1}}{|A|} \\ \frac{A_{12}}{|A|} & \frac{A_{22}}{|A|} & \cdots & \frac{A_{n2}}{|A|} \\ \vdots & \vdots & & \vdots \\ \frac{A_{1n}}{|A|} & \frac{A_{2n}}{|A|} & \cdots & \frac{A_{nn}}{|A|} \end{bmatrix}$$

Proof

By **theorem 3.12** $A\ (adj\ A) = |A|l_n,\ so\ if\ |A| \neq 0,\ then$

$$A\left(\frac{1}{|A|}(adj\ A)\right) = \frac{1}{|A|}\big(A(adj\ A)\big) = \frac{1}{|A|}(|A|l_n) = l_n$$

Hence

$$A^{-1} = \frac{1}{|A|}(adj\ A)$$

Example 4. Again, consider the matrix of example 2. Then $|A| = -94, and$

$$A^{-1} = \frac{1}{|A|}(adj\ A) = \begin{bmatrix} \frac{18}{94} & \frac{6}{94} & \frac{10}{94} \\ -\frac{17}{94} & \frac{10}{94} & \frac{1}{94} \\ \frac{6}{94} & \frac{2}{94} & -\frac{28}{94} \end{bmatrix}$$

3.5 Other Applications of Determinants

We can use the results developed in theorem 3.12 to obtain another method for solving a linear system of n equations in n unknowns. This method is known as Cramer's Rule.

Theorem 3.13 Cramer's Rule (Gabriel Cramer)

Let

$$a_{11}x_1 + a_{12}x_2 + \cdots + a_{1n}x_n = b_1$$
$$a_{21}x_1 + a_{22}x_2 + \cdots + a_{2n}x_n = b_2$$
$$\vdots$$
$$a_{n1}x_1 + a_{n2}x_2 + \cdots + a_{nn}x_n = b_n$$

be a linear system of n equations in n unknowns, and let $A = [a_{ij}]$ be the coefficient matrix so that we can write the given system as $Ax = b$, where $b = \begin{bmatrix} b_1 \\ b_2 \\ \vdots \\ b_n \end{bmatrix}$

If $|A| \neq 0,$ then the system has the unique solution $x_1 = \frac{|A_1|}{|A|}, x_1 = \frac{|A_2|}{|A|}, \cdots, x_n = \frac{|A_n|}{|A|}$ where A_i is the matrix obtained from A by replacing the ith column of A by b.

Proof

If, $|A| \neq 0$, then, by Theorem 3.8, A is nonsingular. Hence

$$X = \begin{bmatrix} x_1 \\ x_2 \\ \vdots \\ x_n \end{bmatrix} = A^{-1}b = \begin{bmatrix} \frac{A_{11}}{|A|} & \frac{A_{21}}{|A|} & \cdots & \frac{A_{n1}}{|A|} \\ \frac{A_{12}}{|A|} & \frac{A_{22}}{|A|} & \cdots & \frac{A_{n2}}{|A|} \\ \vdots & \vdots & & \vdots \\ \frac{A_{1i}}{|A|} & \frac{A_{2i}}{|A|} & \cdots & \frac{A_{ni}}{|A|} \\ \vdots & \vdots & & \vdots \\ \frac{A_{1n}}{|A|} & \frac{A_{2n}}{|A|} & \cdots & \frac{A_{nn}}{|A|} \end{bmatrix} \begin{bmatrix} b_1 \\ b_2 \\ \vdots \\ b_n \end{bmatrix}$$

This mean that $x_1 = \frac{|A_{1i}|}{|A|} b_1 + \frac{|A_{2i}|}{|A|} + \cdots + \frac{|A_{ni}|}{|A|} b_n \; for \; i - 1,2, \dots, n.$

Now let $A_i = \begin{bmatrix} a_{11} & a_{12} & \cdots & a_{1i-1} & b_1 & a_{1i+1} & \cdots & a_{n1} \\ a_{21} & a_{22} & \cdots & a_{2i-1} & b_2 & a_{2i+1} & \cdots & a_{n2} \\ \vdots & \vdots & \ddots & \vdots & \vdots & \vdots & \ddots & \vdots \\ a_{n1} & a_{n2} & \cdots & a_{ni-1} & b_n & a_{ni+1} & \cdots & a_{nn} \end{bmatrix}$

If we evaluate $|A_i|$ by expanding along the cofactors of the ith column, we find that

$$|A_i| = A_{1i}b_1 + A_{2i}b_2 + \cdots + A_{ni}b_n$$

Hence

$$X_i = \frac{|A_i|}{|A|} \; for \; i = 1,2 \; \cdots, n$$

In the expression for x_i given in equation (1), the determinant, $|A_i| \; of \; A_i$ can be calculated by any method desired. It was only in the derivation of the expression for x_i that we had to evaluate $|A_i|$ by expanding along the ith column.

Example 1. Consider the following linear system:

$$-2x_1 + 3x_2 - x_3 = 1$$

$$x_1 + 2x_2 - x_3 = 4$$

$$-2x_1 - x_2 + x_3 = -3$$

Solution

$|A| = \begin{bmatrix} -2 & 3 & -1 \\ 1 & 2 & -1 \\ -2 & 1 & 1 \end{bmatrix} = -2$ then

$X_1 = \frac{\begin{vmatrix} 1 & 3 & -1 \\ 4 & 2 & -1 \\ -3 & -1 & 1 \end{vmatrix}}{|A|} = \frac{-4}{-2} = 2, X_2 = \frac{\begin{vmatrix} -2 & 1 & -1 \\ 1 & 4 & -1 \\ -2 & -3 & 1 \end{vmatrix}}{|A|} = \frac{-6}{-2} = 3$ and

$X_3 = \frac{\begin{vmatrix} -2 & 3 & 1 \\ 1 & 2 & 4 \\ -2 & -1 & -3 \end{vmatrix}}{|A|} = \frac{-8}{-2} = 4$

Note that at this point we have shown that the following statements are equivalent for an n x n matrix a:

1. A is nonsingular.

2. Ax= 0 has only the trivial solution

3. A is row (column) equivalent to $l_{n.}$

4. The linear system $A_x =$
$b\ has\ a\ unique\ solution\ for\ every\ n\ x\ 1\ matrix\ b.$

5. A is a product of elementary matrices.

6. $|A| \neq 0$

EXERCISES

1. Evaluate: $\begin{bmatrix} 3 & 2 \\ -1 & 4 \end{bmatrix}$

2. Evaluate $\begin{vmatrix} 2 & 0 & 0 \\ 0 & 5 & 0 \\ 0 & 0 & -1 \end{vmatrix}$

3. If possible, solve the following linear system by Cramer's rule:

$$3x_1 + 4x_2 + 6x_3 = 4$$

$$X_1 + 2x_2 \qquad = 4$$

$$X_1 + 3x_2 - x_3 = 8$$

4. Repeat exercise 3 for the mean system

$$\begin{bmatrix} 1 & 1 & 1 & -2 \\ 0 & 2 & 1 & 3 \\ 2 & 1 & -1 & 2 \\ 1 & -1 & 0 & 1 \end{bmatrix}\begin{bmatrix} x_1 \\ x_2 \\ x_3 \\ x_4 \end{bmatrix} = \begin{bmatrix} -4 \\ 4 \\ 5 \\ 4 \end{bmatrix}$$

Use the method given in Corollary 3.4, to find the inverse, if it exist, of

5. $\begin{bmatrix} 1 & 1 & 2 \\ 2 & 1 & 2 \\ 3 & 2 & 0 \end{bmatrix}$

6. $\begin{bmatrix} 1 & 1 & 1 & 1 \\ 1 & 2 & -1 & 2 \\ 1 & -1 & 2 & 1 \\ 1 & 3 & 3 & 2 \end{bmatrix}$

7. $\begin{bmatrix} 1 & 1 & 1 & 1 \\ 1 & 3 & 1 & 2 \\ 1 & 2 & -1 & 1 \\ 5 & 9 & 1 & 6 \end{bmatrix}$

8. $\begin{bmatrix} 1 & 1 & 1 \\ 0 & 2 & 3 \\ 1 & 1 & 2 \end{bmatrix}$

Solve the following linear systems by Cramer's rule.

9.
$$x_1 - 2x_2 + 3x_3 = 0$$
$$x_1 - 3x_2 + 2x_3 = 0$$
$$4x_1 + 2x_2 + x_3 = 0$$

10.
$$x_1 + 2x_2 + 4x_3 = 0$$
$$-2x_1 \qquad - 4x_3 = 0$$
$$4x_2 + 4x_3 = 0$$

ASSESSMENT

A. Evaluate the determinants.

1. $\begin{vmatrix} 2 & 0 & 0 & 0 \\ -5 & 3 & 0 & 0 \\ 3 & 2 & 4 & 0 \\ 4 & 2 & 1 & -5 \end{vmatrix}$ 2. $\begin{vmatrix} 4 & 2 & 2 & 0 \\ 2 & 0 & 0 & 0 \\ 3 & 0 & 0 & 1 \\ 0 & 0 & 1 & 0 \end{vmatrix}$ 3. $\begin{vmatrix} -4 & 2 & 0 & 0 \\ 2 & 3 & 1 & 0 \\ 3 & 1 & 0 & 2 \\ 1 & 3 & 0 & 3 \end{vmatrix}$

B. Compute all the cofactors of the following.

4. $\begin{vmatrix} 1 & 2 & -3 \\ 3 & 1 & 4 \\ 4 & 3 & -2 \end{vmatrix}$ 5. $\begin{vmatrix} -1 & 2 & 3 \\ -2 & 5 & 4 \\ 0 & 1 & -3 \end{vmatrix}$ 6. $\begin{vmatrix} 1 & 2 & -3 \\ 3 & 1 & 4 \\ 4 & 3 & -2 \end{vmatrix}$

C. Solve the following linear system by Cramer's Rule

7.
$$x_1 - 2x_2 + 3x_3 = 5$$
$$x_1 - 3x_2 + 2x_3 = 5$$
$$2x_2 + x_3 = 3$$

8.
$$x - y + z = -1$$
$$2x + y - 3z = 8$$
$$x - 2y + 3z = -5$$

Solve the following linear systems by Cramer's rule:

9.
$$x - y + z = -1$$
$$2x + y - 3z = 8$$
$$x - 2y + 3z = -5$$

10.
$$x + 2y + 3z = 1$$
$$x - 2y - z = -1$$
$$3x + y + z = 3$$

KEY TO CORRECTION

<table>
<tr><th>Exercise</th><th>Assessment</th></tr>
<tr><td>1. 14
3. $x_1 = -4, x_2 = 4, x_3 = 0$

5. $\begin{vmatrix} -1 & 1 & 0 \\ \frac{3}{2} & -\frac{3}{2} & \frac{1}{2} \\ \frac{1}{4} & \frac{1}{4} & -\frac{1}{4} \end{vmatrix}$

7. Since row consists solely of zeros then the determinant of the matrix is equals 0.

9. $x = 0, y = 0\ and\ z = 0$</td><td>1. -120

3. 2

5. $\begin{vmatrix} -19 & -6 & -2 \\ 9 & 3 & 1 \\ -7 & -2 & -1 \end{vmatrix}$

7. $x_1 = 12, x_2 = 2, x_3 = -1$

9. $x = 2, y = -1, z = 0$</td></tr>
</table>

Chapter 4

REAL VECTOR SPACES

OBJECTIVES

After working through this section, you should be able to:

- Introduce the basic concepts of vector spaces.
- Define addition of vectors and scalar multiplication
- Describe three examples of real vector space.
- Develop logical thinking and the ability
- Analyze mathematical arguments.
- Enhance your problem-solving abilities.
- Learn the definition of a subspace.
- Learn to determine whether or not a subset is a subspace.
- Learn the most important examples of subspaces.
- Learn to write a given subspace as a column space or null space.
- Determine if a subset of a vector space is a subspace.
- Describe in your own words how lines and planes can be used to visualizesubspaces of $R^{3\times 1}$.
- Utilize the subspace test to determine if a set is a subspace of a given vector space

VECTOR SPACES

Sourchttps://www.google.com/search?q=VECTOR+spaces+pictures&tbm=isch&ved=2ahUKE
wiavOGn2obwAhUFXJQKHRGZBAoQ2cCegQIABAA&oq=VECTOR+spaces+pictures&gs_lcp=

4.1 Vectors Spaces

INTRODUCTION

In various parts of mathematics, one is confronted with a set, such that it is both meaningful and interesting to deal with 'linear combinations' of the objects in that set.For example, in our study of linear equations, we found it quite natural to consider linear combinations of the rows of a matrix. It is likely that the reader has studied calculus and dealt with linear combinations of functions; certainly, this is so if he hasstudied differential equations. Perhaps the reader has had some experience with vectors in three-dimensional Euclidean space, particularly with linear combinations of such vectors.[1]

Loosely speaking, linear algebra is that branch of mathematics that treats the common properties of algebraic systems, which consist of a set, together with a reasonable notion of a 'linear combination' of elements in the set. In this section, we shall define the mathematical object which experience has shown to be the most useful abstraction of this type of algebraic system.[2]

DISCUSSION

Definition. A vector space (or linear space) consists of the following:[3]

1. a field F of scalars;
2. a set V of objects, called vectors;
3. a rule (or operation), called vector addition, which associates with each pair ofvectors $\propto, \beta$ in V a vector $\propto + \beta$ in V, called the sum of $\propto$ *and* β in such a way that

   (a) addition is commutative, $\alpha + \beta = \beta + \alpha$;

   (b) addition is associative, $\alpha + (\beta + \gamma) = (\alpha + \beta) + \gamma$;

   (c) there is a unique vector 0 in V, called the zero vector, such that $\alpha + 0 = \alpha$for all α in V;

   (d) for each vector α in V there is α unique vector $-\alpha$ in V such that $\alpha + (-\alpha) = 0$;

4. a rule (or operation), called scalar multiplication, which associates with each scalar in F and vector α in V a vector $C\alpha$ in V, called the product of c and α, in such a way that [4]

   (a) $l\alpha = \alpha$ for every α in V ;
   (b) $(clc2)\ \alpha = cl\ (c2\alpha)$;
   (c) (c) $c(\alpha + \beta) = c\alpha + c\beta$;
   (d) (d) $(cl + c2)\ \alpha = cl\alpha + c2\alpha$.

It is important to observe, as the definition states, that a vector space is a composite object consisting of a field, a set of 'vectors,' and two operations with certainspecial properties. The same set of vectors may be part of several distinct vector spaces (see Example 5 below). When there is no chance of confusion, we may simplyrefer to the vector space as V, or when it is desirable to specify the field, we shall sayV is a **vector space over the field *F***. The name 'vector' is largely applied to the set V elements largely as a matter of convenience. The origin of the name is to be found in Example 1 below, but one should not attach too much significance to the name, sincethe variety of objects occurring as the vectors in V may not bear much resemblance to any preassigned concept of vector which the reader has. We shall try to indicate this variety by a list of examples; our list will be enlarged considerably as we study vector spaces.[2]

EXAMPLE 1. The n-tuple space, F^n. Let F be any field, and let V be the set ofall n-tuples α = (xl, x2, ... , xn) of scalars xi in F. If β = (yl, y2, . . . , yn) with yi in F, the sum of α *and* β is defined by

(2-1) α *and* β $= x_1 + y_1, x_2 + y_2 \ldots, x_n + y_n$

The product of a scalar c and vector a is defined by

(2-2) $c\alpha = (cx_1, cx_2, \ldots\ldots, cx_n)$

The fact that this vector addition and scalar multiplication satisfy conditions (3)and (4) is easy to verify, using the similar properties of addition and multiplication of elements of F.[1]

EXAMPLE 2. The space of m x n matrices, $F^{mx\,n}$. Let F be any field and let mand n be positive integers. Let F^{mxn} be the set of all m x n matrices over the field F. The sum of two vectors A and B in F^{mxn} is defined by[1]

(2-3) $(A + B)_{ij} = A_{ij} + B_{ij}$

The product of a scalar c and the matrix A is defined by

(2-4) $(cA)_{ij} = cA_{ij}$

$Note\ that\ F^{1xn} = F^n$

EXAMPLE 3. The space of functions from a set to a field. Let F be any fieldand let S be any non-empty set. Let V be the set of all functions from the set S into F.The sum of two vectors f and g in V is the vector f + g, i.e., the function from S into F,defined by[1]

$$(2-5) \qquad (f+g)(s) = f(s) + g(s)$$

The product of the scalar c and the function f is the function cf defined by [2]

$$(2-6) \qquad (cf)(s) = cf(s)$$

The preceding examples are special cases of this one. For an n-tuple of elements of F may be regarded as a function from the set S of integers 1, . .. , n into F. Similarly,an m X n matrix over the field F is a function from the set S of pairs of integers, (i, j), $1 \leq i \leq m$, $1 \leq j \leq n$, into the field F. For this third example, we shall indicate how one verifies that the operations we have defined satisfy conditions (3) and (4). For vector addition:[2]

(a) Since addition in F is commutative,

f(s) + g(s) = g(s) + f(s)

for each s in S, so the functions f + g and g + f are identical.

(b) Since addition in F is associative,

f(s) + [g(s) + h(s)] [f(s) + g(s)] + h(s)

for each s, so f + (g + h) is the same function as (f + g) + h.

(c) The unique zero vector is the zero function which assigns to each element of S the scalar 0 in F.

(d) For each f in V, (-f) is the function which is given by (-f) (s) - f(s).

The reader should find it easy to verify that scalar multiplication satisfies theconditions of (4), by arguing as we did with the vector addition[1]

EXAMPLE 4. The space of polynomial functions over a field F. Let F be a field and let V be the set of all functions f from F into F which have a rule of form [2]

(2-7) f(x) Co + CiX + . .. + cnx^n

where Co, Cl, Cn are fixed scalars in F (independent of x). A function of this type is called a **polynomial function** on F. Let addition and scalar multiplication are defined in Example 3. If j and g are polynomial functions and C is in F, then f + g and cf are again polynomial functions.[1]

EXAMPLE 5. The field C of complex numbers may be regarded as a vector space over the field It of real numbers. More generally, let P be the field of real numbers and let V be the set of n-tuples a (XI, , Xn) where XI, ..., Xn are complex numbers. Define addition of vectors and scalar multiplication by (2-1) and (2-2), as inExample 1. In this way, we obtain a vector space over the field R, which is quite different from the space Cn and the space Rn.[5]

A few simple facts follow almost immediately from the definitionof a vector space, and we proceed to derive these. If c is a scalar and 0 is the zero vector, then by 3(c) and 4(c). [2]

$$cO = c(O + 0) = cO + cO.$$

Adding - (cO) and using 3(d), we obtain

(2-8) $$cO = O.$$

Similarly, for the scalar 0 and any vector, we find that

(2-9) $$O\alpha = O.$$

If c is a non-zero scalar and α is a vector such that $c\alpha = 0$, then by (2-8), c-1(ca) O. But

$$c\text{-}1(c\alpha) = (c\text{-}1c)\, \alpha = 1\alpha = \alpha$$

hence, α = O. Thus we see that if c is a scalar and α a vector such that $c\alpha = 0$, then either c is the zero scalar or α is the zero vector.

If a is any vector in V, then

$$0=0\alpha = (1 - 1)\alpha = 1\alpha + (-1)\, \alpha = \alpha + (-1)\, \alpha$$

from which it follows that

(2-10) $$(-1)\, \alpha = -\alpha$$

Finally, the associative and commutative properties of vector addition imply that a sum involving some vectors is independent of how these vectors are combined and associated. For example, if a_1, a_2, a_3, a_4 are vectors in V, then [2]

$$(\alpha_1 + \alpha_2) + (\alpha_3 + \alpha_4) = [\alpha_2 + (\alpha_1 + \alpha_2)] + \alpha_4$$

and such a sum may be written without confusion as

$$a_1 + a_2 + a_3 + a_4$$

Definition. A vector β in V is said to be a **linear combination** of the vectors $a_1, \cdots, a_n$ in V provided there exist scalars $c_1, \cdots, c_n$ in F such that

$$\beta = c_1\alpha_1 + \cdots + C_n\alpha_n$$

$$= \sum_{i=1}^{n} c_i\alpha_i$$

Other extensions of the associative property of vector addition and the distributiveproperties 4(c) and 4(d) of scalar multiplication apply to linear combinations:[2]

$$\sum_{i=1}^{n} c_i\alpha_i + \sum_{i=1}^{n} d_i\alpha_i = \sum_{i=0}^{n} (c_i + d_i)\alpha_i$$

$$c\sum_{i=1}^{n} c_i\alpha_i = \sum_{i=0}^{n} (cc_i)$$

Certain parts of linear algebra are intimately related to geometry. The very word'space' suggests something geometrical, as does the word 'vector' to most people. Aswe proceed with our study of vector spaces, the reader will observe that much of the terminology has a geometrical connotation. Before concluding this introductory section on vector spaces, we shall consider the relation of vector spaces to geometry to an extent that will at least indicate the origin of the name 'vector space.' This will be a brief intuitive discussion.[1]

Let us consider the vector space R3. In analytic geometry, one identifies triples $(x_1, x_2, x_3)$of real numbers with the points in three-dimensional Euclidean space. In that context, a vector is usually defined as a directed line segment PQ, from a point P in the space to another point Q. This amounts to a careful formulation of the idea of the 'arrow' from P to Q. As vectors are used, it is intended that they should be determined by their length and direction. Thus, one must identify two directed line segments if they have the same length and the same direction[1]

The directed line segment PQ, from the point P = (x_1, x_2, x_3) to the point Q = (y_1, y_2, y_3) has the same length and direction as the directed line segment from the origin 0 = (0, 0, 0) to the point $(y_1 - x_1, y_2 - x_2, y_3 - x_3)$. Furthermore, this is the only segment emanating from the origin which has the same length and direction as PQ. Thus, if one agrees to treat only vectors which emanate from the origin, there is exactly one vector associated with each given length and direction.[2]

The vector OP, from the origin to $P = (x_1, x_2, x_3)$ is completely determined by P, and it is possible to identify this vector with the point P. In our definition of the vector space R3, the vectors are simply defined to be the triples (x_1, x_2, x_3).[1]

Given points $P = (x_1, x_2, x_3)$ and $Q = (y_1, y_2, y_3)$, the definition of the sum of the vectors OP and OQ can be given geometrically. If the vectors are not parallel, then the segments OP and OQ determine a plane and these segments are two of the edges of a parallelogram in that plane (see Figure 1). One diagonal of this parallelogram extends from 0 to a point S, and the sum of OP and OQ is defined to be the vector OS. The coordinates of the point S are $(x_1 + y_1, x_2 + y_2, x_3 + y_3)$ and hence this geometrical definition of vector addition is equivalent to the algebraic definition of Example 1.[2]

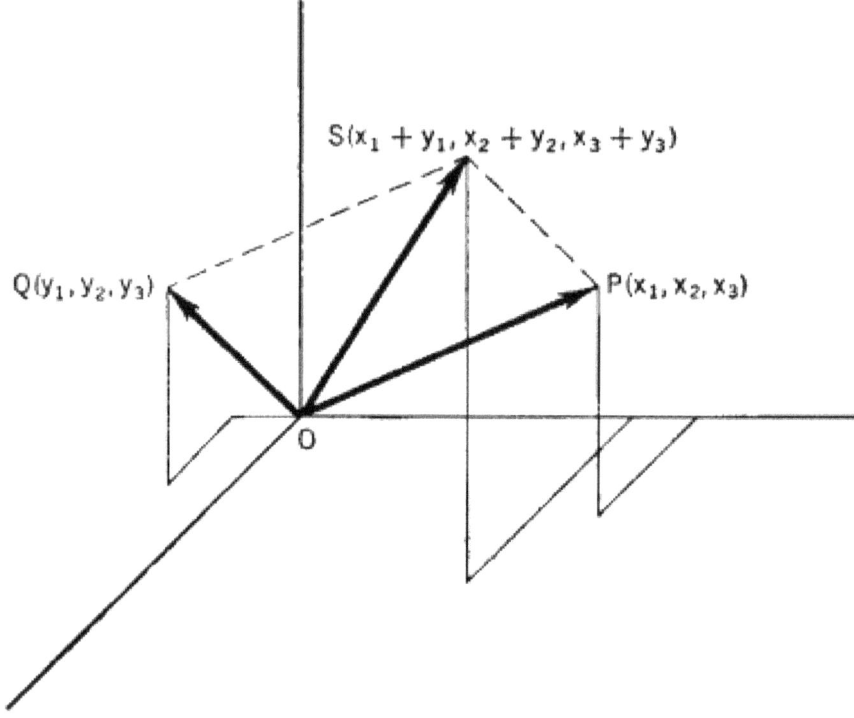

FIGURE 1

Scalar multiplication has a simpler geometric interpretation. If c is a real number, then the product of e and the vector OP is the vector from the origin with length lei times the length of OP and a direction which agrees with the direction of OPif c > 0, and which is opposite to the direction of OP if c < 0. This scalar multiplicationjust yields the vector OT where T (cx_1, cx_2, cx_3), and is therefore consistent with the algebraic definition given for R^3.[6]

From time to time, the reader will probably find it helpful to 'think geometrically' about vector spaces, that is, to draw pictures for his own benefit to illustrate and motivate some of the ideas. Indeed, he should do this. However, in forming such illustrations, he must bear in mind that, because we are dealing with vector spaces as algebraic systems, all proofs we give will be of an algebraic nature.[2]

EXERCISES

1. Is the set of ordered pairs V = { (x_1 , x_2) | $x_1 \in R$, $x_2 \in R$ } a vector space?

2. Let V be a vector space over field R and u, v ∈ V. Simplify each of the following:
   (i) 4(5u−6v) +2(3u+v)
   (ii) 6(3u+2v) +5u−7v
   (iii) 5(2u−3v) +4(7v+8)
   (iv) 3(5u+ 2/v)

3. Mark each of the following as true or false.

   (i) Null vector in a vector space is unique.

   (ii) Let V(F) be a vector space. Then scalar multiplication on V is abinary operation on V.

   (iii) Let V(F) be a vector space. Then au = av ⇒ u = v for all u, v ∈ Vand for all α ∈ F.

   (iv) Let V(F) be a vector space. Then au = bu ⇒ a = b for all u ∈ Vand for all α, b ∈ F.

KEY TO CORRECTION

1. Until operations of vector addition and scalar multiplication are specified, we cannot test for vector space structure. Therefore, we do not have a vector space.

2. (i) $26u-22v$

   (ii) $23u+5v$

   (iii) The sum $7v+8$ is not defined, so the given expression is not meaningful.

   (iv) Division by v is not defined, so the given expression is not meaningful.

3.

   (i) T
   (ii) F
   (iii) F
   (iv) F

1. If F is a field, verify that Fn (as defined in Example 1) is a vector space over the field F.

2. If V is a vector space over the field F, verify that

   $(\alpha 1 + \alpha 2) + (\alpha 3 + \alpha 4) = [[\alpha 2 + (\alpha 3 + \alpha 1)] + a4$ for all vectors $\alpha 1$, $\alpha 2$, $\alpha 3$, and $\alpha 4$ in V.

3. If C is the field of complex numbers, which vectors in C3 are linear combinations of (1, 0, - 1), (0, I, 1), and (1, 1, I)?

4. Let V be the set of all pairs (x, y) of real numbers, and let F be the field of realnumbers. Define

   (x, y) + (xl, yl) = (x + xl, y + yl)

   c(x, y) = (cx, y).

   Is V, with these operations, a vector space over the field of real numbers?

5. On Rn, define two operations

$$\alpha\ o = \alpha - \beta$$
$$c.a = -a$$

The operations on the right are the usual ones. Which of the axioms for a vector space are satisfied by $(R^n, \oplus,)$?

6. Let V be the set of all complex-valued functions f on the real line such that (for all t in R)

$$f(-t) = f(t)$$

The bar denotes complex conjugation. Show that V, with the operations

(f + g) (t) = f(t) + g(t)

(cf)(t) = cf(t)

is a vector space over the field of real numbers. Give an example of a function in Vwhich is not real-valued.

7. Let V be the set of pairs (x, y) of real numbers and let F be the field of real numbers. Define

$$(x, y) + (x_1, y_1) = (Xx + x_1, 0)$$

$$c(x, y) = (cx, 0).$$

Is V, with these operations, a vector space?

SUBSPACE

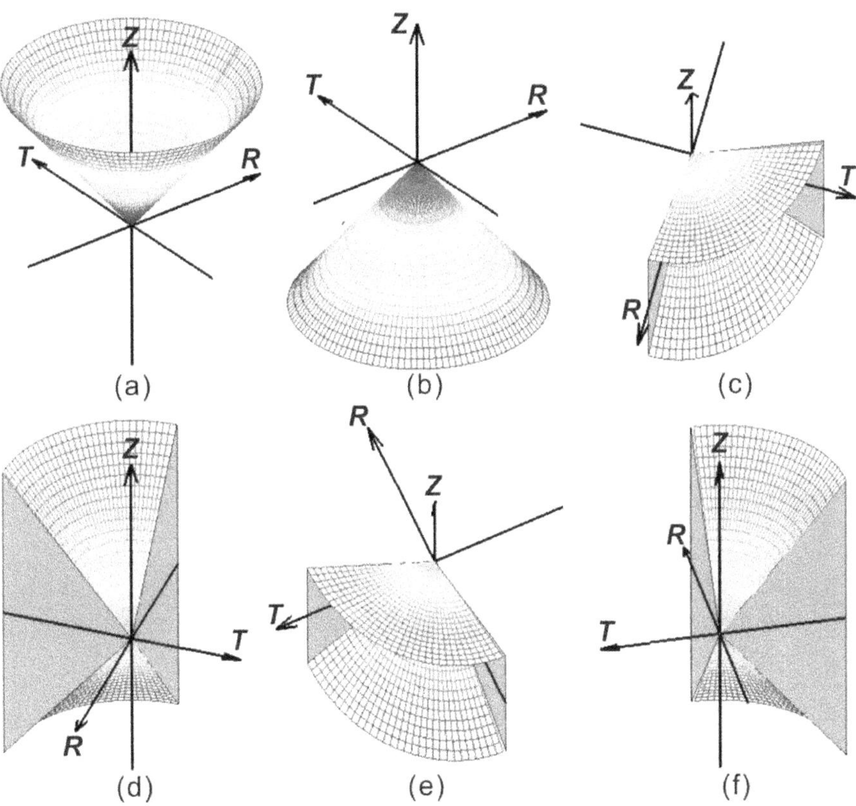

https://www.google.com/search?q=subspace&tbm=isch&ved=2ahUKEwjroIf hrZHwAhULJqY KHW1IBRwQ2-cCegQIABAA&oq=subspace&gs_lcp

4.2 Subspace

INTRODUCTION

In this section, we shall introduce some of the basic concepts in the study of vector spaces.

Recall that a vector space consists of a set V along with two operations, addition, and scalar multiplication. Moreover, the addition operation satisfies the abelian-group axioms, while the multiplication satisfies the scalar axioms. Now suppose that $W \subseteq V$ is a subset of V. Then the addition and multiplication operations are also defined over W. Does this make W a vector space? Not necessarily. The following are why W can fall short of being a vector space. In what follows, we'll assume that $V = R^2$

DISCUSSION

Definition. Let V be a vector space over the field F. A **subspace** of V is a subset W of V, which is itself a vector space over F with the operations of vector addition and scalar multiplication on V.[3]

A direct check of the axioms for a vector space shows that the subset W of V is a subspace if for each α and β in W the vector $\alpha + \beta$ is again in W; the 0 vector is in W; for each α in W the vector $(-\alpha)$ is in W; for each α in W and each scalar c, the vector c α is in W. The commutativity and associativity of vector addition and the properties (4) (a), (b), (c), and (d) of scalar multiplication do not need to be checked, since these are properties of the operations on V. One can simplify things still further.[2]

Theorem 1. A non-empty subset W of V is a subspace of V if and only if for each pairof vectors α, β in W and each scalar c in F the vector c α + β is again in W.

Proof. Suppose that W is a non-empty subset of V such that c α + β belongs toW for all vectors α,β in W and all scalars c in F. Since W is non-empty, there is a vector ρ in W, and hence (- l) ρ + ρ = 0 is in W. Then if α is any vector in W and c anyscalar, the vector cα = c α + 0 is in W. In particular, (-1) α= - α is in W. Finally, if α and β are in W, then α + β = 1 α + β is in W. Thus, W is a subspace of V.[3]

Conversely, if W is a subspace of V, α and β are in W, and c is a scalar; certainly, $c\alpha$ + β is in W.

Some people prefer to use the $c\alpha$ + β property in Theorem 1 as the definition of a subspace. It makes little difference. The important point is that, if W is a non-empty subset of V such that $c\alpha$ + β is in V for all α , β in W and all c in F, then (with the operations inherited from V) W is a vector space. This provides us with many new examples of vector spaces.[2]

EXAMPLE 6.

(a) If V is any vector space, V is a subspace of V; the subset consisting of thezero vector alone is a subspace of V, called the **zero subspace** of V.[1]

(b) In F^n, the set of n-tuples (xl, ... , xn) with x1 = 0 is a subspace ; however, the set of n-tuples with x1=1 + x2 is not a subspace (n ≥ 2).[7]

(c) The space of polynomial functions over the field F is a subspace of the spaceof all functions from F into F.[2]

(d) An n X n (square) matrix A over the field F is **symmetric** if

Aij= Aji for each i and j. The symmetric matrices form a subspace of the space of all nX n matrices over F.[1]

(e) An n X n (square) matrix A over the field C of complex numbers is **Hermitian** (or **self-adjoint**) if

$$A_{ij} = A_k$$

for each j, k, the bar denoting complex conjugation. A 2 X 2 matrix is Hermitian if and only if it has the form. [6]

$$\begin{vmatrix} z & x+iy \\ x-iy & w \end{vmatrix}$$

where x, y, z, and w are real numbers. The set of all Hermitian matrices is *not* a subspace of the space of all n X n matrices over C. For if A is Hermitian, its diagonal entries A11, A22) ... , are all real numbers, but the diagonal entries of iA are generally not real. On the other hand, it is easily verified that the set of n X n complex Hermitian matrices is a vector space over the field R of real numbers (with the usual operations).[2]

EXAMPLE 7. The solution space of a system of homogeneous linear equations. Let A be an m X n matrix over F. Then the set of all n X 1 (column) matricesX over F such that AX = 0 is a subspace of the space of all n X 1 matrix over F. To prove this, we must show that A (cX + Y) = 0 when AX = 0, A Y = 0, and c is an arbitraryscalar in F. This follows immediately from the following general fact.[6]

Lemma. If A is an m X n matrix over F and B, C are n X p matrices over F then

(2-11) A (dB + C) = d(AB) + AC

for each scalar d in F.

$$Proof.[A(dB+C)]_{ij} = \sum_k A_{ik}(dB+C)_{kj}$$

$$= \sum_k (dA_{ik}B_{kj} + A_{ik}C_{kj})$$

$$= d\sum_k A_{ik}B_{kj} + \sum_k A_{ik}\,C_{kj}$$

$$= d\,(AB)_{ij} + (AC)_{ij}$$

$$= [d(AB) + AC]_{ij}$$

Similarly, one can show that (dB + C) A = d(BA) + CA, if the matrix sums and productsare defined.[8]

Theorem 2. Let V be a vector space over the field F. The intersection of any collection of subspaces of V is a subspace of V.[1]

Proof. Let {Wa} be a collection of subspaces of V, and let W = ∩ Wa be their intersection. Recall that W is defined as the set of all elements belonging to every Wa (see Appendix). Since each Wa is a subspace, each contains the zero vector. Thus, the zero vector is in the intersection W, and W is non-empty. Let α and β be vectors in W and let c be a scalar. By definition of W, both α and β belong to each Wa, and because each Wa is a subspace, the vector (cα and β) is in every Wa. Thus ($c\alpha$+β) isagain in W. By Theorem 1, W is a subspace of V.[3]

From Theorem 2 it follows that if S is any collection of vectors in V, then there is a smallest subspace of V which contains S, that is, a subspace that contains S and which is contained in every other subspace containing S.[1]

Definition. Let S be a set of vectors in a vector space V. The **subspace spanned** by S is defined to be the intersection W of all subspaces of V which contain S. When S is a finite set of vectors, S = {a1, a2, ... , an}, we shall simply call W the **subspace spanned by the vectors** a1, a2, ... , an.

Theorem 3. The subspace spanned by a non-empty subset S of a vector spaceV is the set of all linear combinations of vectors in S.[6]

Proof. Let W be the subspace spanned by S. Then, each linear combination

$$\alpha = x_1 \alpha_1 + x_2 \alpha_2 \ ...+ x_m \alpha_m$$

of vectors α1, α2, ..., αm in S is clearly in W. Thus, W contains the set L of all linear combinations of vectors in S. The set L, on the other hand, contains S and is non-empty. If α , β belong to L, then

α is a linear combination,[2]

$$\alpha = x_1\alpha_1 + x_2\alpha_2 + \ldots + x_m\alpha_m$$

of vectors $\alpha 1$ in S, and β is a linear combination,

$$\beta = y_1\beta_1 + y_2\beta_2 + \ldots + y_n\beta_n$$

of vectors (3j in S. For each scalar c,

$$c\alpha + \beta = \sum_{i=1}^{m}(cx_i)\alpha_i + \sum_{j=1}^{n} y_i\beta_j$$

Hence cα + β belongs to L. Thus, L is a subspace of V.

Now we have shown that L is a subspace of V that contains S, and any subspace that contains S contains L. It follows that L is the intersection of all subspaces containing S, i.e., that L is the subspace spanned by the set S.[2]

Definition. If S1, S2, ..., Sk are subsets of a vector space V, the set of all sums [9]

$$\alpha_1 + \alpha_2 + \ldots + \alpha_k$$

of vectors αi in Si is called the **sum** of the subsets S1, S2, ... , Sk and is denoted by or by [9]

$$S_1 + S_2 + \cdots + S_k \text{ or by}$$

$$\sum_{i=1}^{k} S_i$$

If $W_1, W_2, \cdots, W_k$ are subspaces of V, then then sun

$$W = W_1 + W_2 + \cdots + W_k$$

is easily seen as a subspace of V that contains each of the subspaces Wi. From this it follows, as in the proof of Theorem 3, that W is the subspace spanned by the union of WI, W2, ··· , Wk.[1]

EXAMPLE 8. Let F be a sub field of the field C of complex numbers. Suppose

$$\alpha_1 = (1, 2, 0, 3, 0)$$

$$\alpha_2 = (0, 0, 1, 4, 0)$$

$$\alpha_3 = (0, 0, 0, 0, 1)$$

By Theorem 3, a vector a is in the subspace W of F^s spanned by $\alpha 1$, $\alpha 2$, $\alpha 3$ if and only if there exist scalars cl, c2, c3 in F such that

$$\alpha = c_1\alpha_1 + c_2\alpha_2 + c_3\alpha_3$$

Thus W consists of all vectors of the form

$$\alpha = (c_1, 2c_1, c_{2,3}\, c_1 + 4\, c_2, c_3)$$

where c1, c2, c3 are arbitrary scalars in F. Alternatively, W can be described as the setof all 5-tuples

$$\alpha = (x_1, x_2, x_3, x_4, x_5)$$

with x1 in F such that

$$x_2 = 2x_l$$

$$x_4 = 3x_l + 4x_3$$

Thus (-3, -6, 1, 5, 2) is in W, whereas (2, 4, 6, 7, 8) is not.

EXAMPLE 9. Let F be a subfield of the field C of complex numbers, and let V be the vector space of all 2 X 2 matrices over F. Let Wl be the subset of V consisting of all matrices of the form.[8]

$$\begin{bmatrix} x & y \\ z & 0 \end{bmatrix}$$

where x, y, z are arbitrary scalars in F. Finally, let W2 be the subset of V consisting ofall matrices of the form. [7]

$$\begin{bmatrix} x & 0 \\ 0 & y \end{bmatrix}$$

where x and y are arbitrary scalars in F. Then Wl and W2 are subspaces of V. Also [2]

$$V = W_1 + W_2$$

because

$$\begin{bmatrix} a & b \\ c & d \end{bmatrix} = \begin{bmatrix} a & b \\ c & 0 \end{bmatrix} + \begin{bmatrix} 0 & 0 \\ 0 & d \end{bmatrix}$$

The subspace W1 ∩ W2 consists of all matrices of the form

$$\begin{bmatrix} x & 0 \\ 0 & 0 \end{bmatrix}$$

EXAMPLE 10. Let A be an m X n matrix over a field F. The **row vectors** of A are the vectors in F^n given by αI = (A i1, ..., A in), i = 1, . . . , m. The subspace of Fn spanned by the row vectors of A is called the **row space** of A. The subspace considered in Example 8 is the row space of the matrix. [2]

$$A = \begin{bmatrix} 1 & 2 & 0 & 3 & 0 \\ 0 & 0 & 1 & 4 & 0 \\ -4 & -8 & 1 & -8 & 0 \end{bmatrix}$$

It is also the row space of the matrix

$$B = \begin{bmatrix} 1 & 2 & 0 & 3 & 0 \\ 0 & 0 & 1 & 4 & 0 \\ 0 & 0 & 0 & 0 & 1 \\ -4 & -8 & 1 & -8 & 0 \end{bmatrix}$$

EXAMPLE 11. Let V be the space of all polynomial functions over F. Let S bethe subset of V consisting of the polynomial functions f_0, f1 f2 ,defined by n = 0, 1,2,... Then V is the subspace spanned by the set S. [1]

$$f_n(x) = x^n, \ n = 0, 1, 2, \ldots\ldots$$

Then V is the subspace spanned by the set.

EXERCISES

1. In each part, V is a vector space and S is a subset of V. Determine whether Sis a subspace of V

   (a) $V = R^3$

   $$S = \left\{ \begin{bmatrix} x \\ 12 \\ 3x \end{bmatrix} x \in R \right\}$$

   (b) $V = R^2$

   $$S = \left\{ \begin{bmatrix} x \\ y \end{bmatrix} 2x - 5y = 11 \right\}$$

2. State the conditions for $W \subset V$ to be a subspace of V .

3. Which of the following is a subspace of R3 ?

   a. W1 = {(x1, x2, x3) ∈ R3 | x1 − x2 = 1}
   b. (b) W2 =< (1, 0, 2),(2, 1, 1) >
   c. W3 = {(x1, x2, x3) ∈ R3 | x1 = 0}
   d. W4 = {(x1, x2, x3) ∈ R3 | x2 = √ 2x3}

KEY TO CORRECTION

1. (a) S is not a subspace because the zero vector $\begin{bmatrix}0\\0\\0\end{bmatrix}$ cannot be written in the form $\begin{bmatrix}x\\12\\3x\end{bmatrix}$ for any possible value of x, so $\begin{bmatrix}0\\0\\0\end{bmatrix}$ $\in/S$ and S cannot be a subspace.

   b) No, this is not a substance. After all, the zero vector $\begin{bmatrix}0\\0\\0\end{bmatrix}$ is not in S since $2(0) - 5(0) = 0 \neq 11$.

2. Let V be a vector space over a field F, then $W \subset V$ is a subspace of V if and only if $\forall w_1, w_2 \in w$ and $\alpha_1, \alpha_2 \in F$ the vector $\alpha_1 w_1 + \alpha_2 w_2 \in w$. That is W is closed with respect to the sum of vectors and the product by a scalar of the field.

3. (a) $w_1 = \{(x_1, x_2, x_3) \in R^3 | x_1 - x_2 = 1\}$ is not a subspace since $(0,0,0)$ not $\in W_1$

   (b) $w_2 = < (1,0,2), (2,1,1) >$ is a subspace by definition since it is the space generated by 2 vectors.

   (c) $w_3 = \{(x_1, x_2, x_3) \in R^3 | x_1 = 0\}$ is a subspace: if $w_1 = (x_1, x_2, x_3), w_2 = (y_1, y_2, y_3) \in w_3$ then $x_1 = y_1 = 0$. Let $\alpha_1, a_2 \in R$ be scalars then $\alpha_1 w_1 + \alpha_2 w_2 = (\alpha_1 x_1 + \alpha_2 y_1, \alpha_1 x_2 + \alpha_2 y_2, \alpha_1 x_3 + \alpha_2 y_3 \in w_3$ since $x_1 = y_1 = 0$ implies that $\alpha_1 x_1 + \alpha_2 y_1 = 0$.

   (d) $w_4 = \left\{(x_1, x_2, x_3) \in R^3 | x_2 = \sqrt{2x3}\right\}$ is a subspace: if $w_1 = (x_1, x_2, x_3), w_2 = (y_1, y_2, y_3) \in w_4$ then $x_2 = \sqrt{2x3}$ and $y_2 = \sqrt{2y_3}$. Let $\alpha_1, \alpha_2 \in R$ be scalars then $\alpha_1 w_1 + \alpha_2 w_2 = (\alpha_1 x_1 + \alpha_2 y_1, \alpha_1 x_2 + \alpha_2 y_2, \alpha_1 x_3 + \alpha_2 y_3) \in w_4$ since $x_2 = \sqrt{2x3}$ and $y_2 = \sqrt{2y_3}$ implies that $\alpha_1 x_2 + \alpha_2 y_2 = \alpha_1 \sqrt{2x_3} + \alpha_2 \sqrt{2y_3} = \sqrt{2\alpha_1 x_3} + \alpha_2 y_3$.

1. Which of the following sets of vectors $\alpha = (\alpha, \ldots, \alpha_n)$ in R^n are subspaces of R^n ($n \geq 3$) ?

   (a) all α such that $\alpha l \geq 0$;

   (b) all α such that $\alpha l + 3\alpha 2 = \alpha 3$;

   (c) all α such that $\alpha 2 = \alpha^2$; 1

   (d) all α such that $\alpha 1\ \alpha 2 = 0$;

   (e) all α such that $\alpha 2$ is rational.

2. Let V be the (real) vector space of all functions 1 from R into R. Which of the following sets of functions are subspaces of V?

   (a) all f such that f(x2) = f(x)2;

   (b) all f such that f(0) = f(l) ;

   (c) all f such that f(3) = 1 + f (- 5);

   (d) all f such that f(- l) = 0;

   (e) all f which are continuous.

3. Is the vector (3, - 1, 0, - 1) in the subspace of R5 spanned by the vectors (2, -1, 3, 2), (- 1, 1, 1, -3), and (1, 1, 9, - 5) ?

4. Let W be the set of all (xl, x2, x3, x4, x5.) in R5 which satisfy

$$2x_1 - x_2 + \frac{4}{4}x_3 - x_4 = 0$$
$$x_1 + \frac{2}{3}x_3 - x_5 = 0$$
$$9x_1 - 3x_2 + 6x_3 - 3x_5 = 09$$

   Find a finite set of vectors which spans W.

5. Let F be a field and let n be a positive integer ($n \geq 2$) . Let V be the vector space ofall n X n matrices over F. Which of the following

sets of matrices A in V are subspacesof V?

(a) all invertible A;

(b) all non-invertible A;

(c) all A such that AB = BA, where B is some fixed matrix in V;

(d) all A such that A2 = A.

Chapter 5

INNER PRODUCT

INNER PRODUCT SPACES AND ORTHOGONALLY

INTRODUCTION

After working through this section, you will be able to:

1. define inner product and inner product space;
2. describe the properties of inner product and inner product space;
3. recognize how inner products be represented;
4. know when the two vectors are said to be orthogonal;
5. use Gram-Schmidt Process in constructing an orthogonal vector.

Sometimes, we generalized the linear structure (addition and scalar multiplication) of R^2 and R^3 in defining a vector space and ignored other important features, such as the notions of length and angle. These ideas are embedded in the concept we now investigate, inner products.[1]

REVIEW

Linear algebra is a branch of mathematics that deals with and treats the common properties of algebraic systems, which consist of a set, together with a reasonable notion of a 'linear combination' of elements in the set.[2]

In the previous section, we defined the mathematical objectthat experience has shown to be the most useful abstraction of this algebraic system. The vector spaces Rn are not the only vector spaces. We now give a general definition that includes Rn for all values of n, and R^S for all sets S, and more. This mathematical structure applies to a wide range of real-world problems and allows for a tremendous economy of thought; the idea of a basis for a vector space will drive home the main idea of vector spaces; they are sets with very simple structures.[2, 3]

The two key properties of vectors are that they can be added together and multiplied by scalars. Thus, before giving a rigorous definition of vector spaces, we restate the main idea.[3]

INNER PRODUCT SPACES

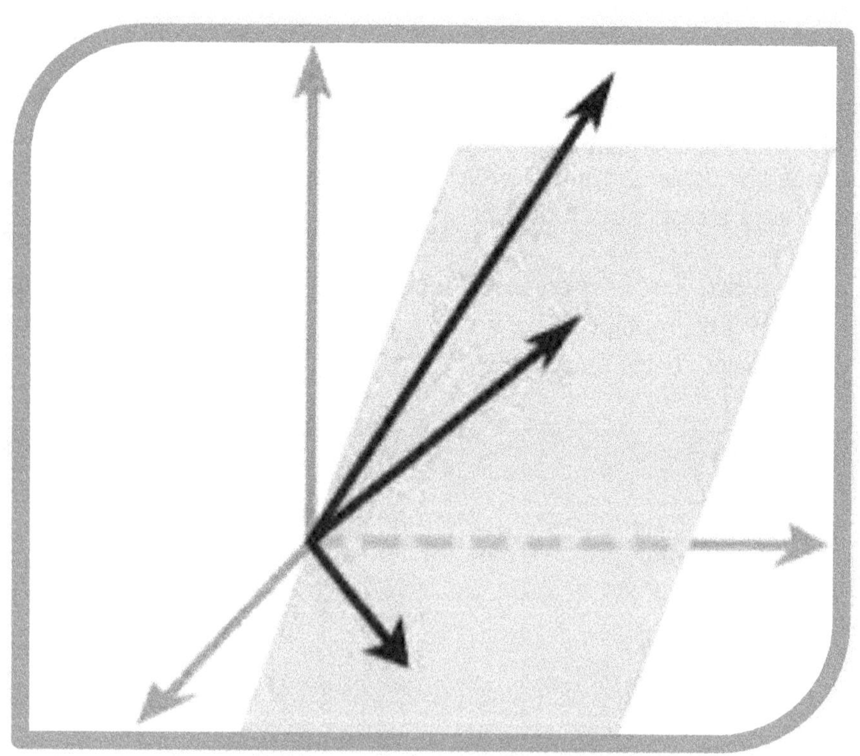

5.1 Inner Product Spaces

DISCUSSION

1. Dot product of R^n

The **inner product or dot product of R^n** is a function ⟨ , defined by [1]

$$\langle u, v\rangle = a_1b_1 + a_2b_2 + \ldots + a_nb_n$$

for $u = [a_1, a_2, \ldots, a_n]^T$, $v = [b_1, b_2, \ldots, b_n]^T \in R^n$

The inner product ⟨ , ⟩ satisfies the following properties:

i. **Linearity:** $\langle au + bv, w\rangle = a\langle u, w\rangle + b\langle v, w\rangle$.

ii. **Symmetric Property**: $\langle u, v\rangle = \langle v, u\rangle$.

iii. **Positive Definite Property**: For any $u \in V$, $\langle u, u\rangle \geq 0$; and

(b) $\langle u, u\rangle = 0$ if and only if $u = 0$.

The vector space V with an inner product is called **a (real) inner product space.**

x_1 y_1

Example 1.1 $x = \begin{bmatrix} x_1 \\ x_2 \end{bmatrix}, y = \begin{bmatrix} y_1 \\ y_2 \end{bmatrix} \in \boldsymbol{R}^2, define$

$$(\boldsymbol{x}, \boldsymbol{y}) = \boldsymbol{2x_1y_1 - x_1y_1 + 5x_2y_2}$$

Then ⟨ , ⟩ is an inner product of R^2. It is easy to see the linearityand the symmetric property. As for the positive definite property,note that

1

$$(x, x) = 2x_1^2y_1 - 2x_1x_2 + 5x_2^2$$

$$= (X_1 + X_2)^2 + (x_1 - 2x_2)^2 \geq 0$$

Moreover, $(x, x) = 0$ if and only if

$$x + x_2 = 0, x_1 - 2x_2 = 0$$

Which implies $x_1 = x_2 = 0, i, e, x = 0$. This inner product on R^2 is different from the dot product of R^2.

For each vector u $u \in V$, the norm (also called the length) of u is defined as the number

$$\|u\| \cdot := \sqrt{\langle u, u \rangle}.$$

If $\|u\| = 1$, we call **u** a **unit vector,** and **u** is said to be **normalized.** For any nonzero vector $v \in V$, we have the unit vector.[1]

$$\hat{v} = \frac{1}{\|v\|} v.$$

This process is called normalizing v.

Let $B = \{u_{1,}u_2 + \because,\because,, u_n\}$ be a basis of an n-dimensional inner product space V. For vectors $u, v \in V$, write [1]

$$u = x_1u_1 + x_2u_2 + \because\because + x_nu_{n.}$$

$$v = y_1u_1 + y_2u_2 + \because\because + y_nu_n$$

The linearity implies

$$\langle u, v \rangle = \langle \sum_{i=1}^{n} x_iu_{i,} \sum_{j=1}^{n} x_ju_j \rangle$$

$$= \sum_{i=1}^{n} \sum_{j=1}^{n} x_iy_i \langle u_iu_j \rangle$$

We call the n x n matrix

$$A = \begin{bmatrix} \langle u_1u_1 \rangle & \langle u_1u_2 \rangle & \cdots & \langle u_1u_n \rangle \\ \langle u_2u_1 \rangle & \langle u_2u_2 \rangle & \cdots & \langle u_2u_n \rangle \\ \vdots & \vdots & & \vdots \\ \langle u_nu_1 \rangle & \langle u_nu_2 \rangle & \cdots & \langle u_nu_n \rangle \end{bmatrix}$$

The matrix of the inner product $\langle , \rangle$ relative to the basis B. Thus, using coordinates vectors

$$[u]B = [x_1, x_2, \dots, x_n]^T,$$

$$[V]B = [y_1, y_2, \dots, y_n]^T$$

Example 1.2. The vector space R^n with the dot product

$$u.v = a_1b_1 + a_2b_2 + \dots + a_nb_n,$$

Where $u = [a_1, a_2, \dots, a_n]^T$, $v = [b_1, b_2, \dots, b_n]^T \in R^n$ is an inner product space. The vector space R^n with this special inner product (dot product) is called the Euclidean n-space, and the dot product is called the standard inner product on R^n.[1]

Example 1.3. The vector space $M_{m,n}$ of all m x n real matrices can be made into an inner product space under the inner product [1]

$$\langle A, B\rangle = tr\,(B^TA), where\ A, B \in M_{m,n}$$

For instance, when m=3, n=2, and for

$$A = \begin{bmatrix} a_{11} & a_{12} \\ a_{21} & a_{22} \\ a_{31} & a_{32} \end{bmatrix}, B = \begin{bmatrix} b_{11} & b_{12} \\ b_{21} & b_{22} \\ b_{31} & b_{32} \end{bmatrix}$$

We have

$$B^TA = \begin{bmatrix} b_{11}a_{11} + b_{21}a_{21} + b_{31}a_{31} & b_{11}a_{12} + b_{21}a_{22} + b_{31}a_{32} \\ b_{12}a_{11} + b_{22}a_{21} + b_{32}a_{31} & b_{12}a_{12} + b_{22}a_{22} + b_{32}a_{32} \end{bmatrix}$$

Thus

$$\langle A, B\rangle = b_{11}a_{11} + b_{21}a_{21} + b_{31}a_{31} + b_{12}a_{12} + b_{22}a_{22} + b_{32}a_{32}$$

$$= \sum_{i=1}^{3}\sum_{j=1}^{3} a_{ij}b_{ij}$$

This means that the inner product space $(M_{3,2}, \langle , \rangle)$ is isomorphic to the Euclidean space $(R^{3x2}, .)$

Example 1.4. The vector space $C[a,b]$ of all real-valued continuous functions on a closed interval $[a,b]$ is an inner product space, whose inner [roduct is defined by [1]

$$\langle f,g\rangle = \int_a^b f(t)g(t)dt, \quad f,g \in C[a,b].$$

2. Inner Product Representation

Let V be an n-dimensional vector space with an inner product ⟨, ⟩, and let A be the matrix of ⟨, ⟩ relative to a basis B. Then for any vectors u, v ∈ V, [1]

$$\langle u,v\rangle = X^T A_y.$$

Where x and y are the coordinate vectors of u and v, respectively,[1]

i.e., $x = [u]B \; and \; y = [v]B.$

Example 2.1 For the inner product of R[3] defined by

$$\langle x,y\rangle = 2x_1y_1 - x_1y_2 - x_2y_1 + 5x_2y_2$$

Where $x = \begin{bmatrix} x_1 \\ x_2 \end{bmatrix}, y = \begin{bmatrix} y_1 \\ y_2 \end{bmatrix} \in R$, its matrix relative to the standard basis $E = \{e_1, e_2\}$ is

$$A = \begin{bmatrix} \langle e_1,e_1\rangle & \langle e_1,e_2\rangle \\ \langle e_2,e_1\rangle & \langle e_2,e_2\rangle \end{bmatrix} = \begin{bmatrix} 2 & -1 \\ -1 & 5 \end{bmatrix}$$

The inner product can be written as

$$\langle x,y\rangle = x^T A_y = [x_1, x_2] \begin{bmatrix} 2 & -1 \\ -1 & 5 \end{bmatrix} \begin{bmatrix} y_1 \\ y_2 \end{bmatrix}$$

We may change variables so the inner product takes a simple form.

For instance, let

$$\begin{cases} x_1 = (^2/_3)x_1' + (^1/_3)x_2' \\ x_2 = (^1/_3)x_1' - (^1/_3)x_2' \end{cases}$$

$$\begin{cases} y_1 = (2/3)y_1' + (1/3)y_2' \\ y_2 = (1/3)y_1' - (1/3)y_2' \end{cases}$$

We have

$$\langle x, y\rangle = 2\left(\frac{2}{3}x_1' + \frac{1}{3}x_2'\right)\left(\frac{2}{3}y_1' + \frac{1}{3}y_2'\right)$$

$$-\left(\frac{2}{3}x_1' + \frac{1}{3}x_2'\right)\left(\frac{1}{3}y_1' - \frac{1}{3}y_2'\right)$$

$$-\left(\frac{1}{3}x_1' - \frac{1}{3}x_2'\right)\left(\frac{1}{3}y_1' - \frac{1}{3}y_2'\right)$$

$$= x_1'y_1' + x_2'y_2' = x'^T y'$$

This is equivalent to choosing a new basis so that the matrix of the inner product relative to the new basis is the identity matrix.[1]

The matrix of the inner product relative to the basis

$$B = \left\{u_1 = \begin{bmatrix} 2/3 \\ 1/3 \end{bmatrix}, u_2 = \begin{bmatrix} 1/3 \\ -1/3 \end{bmatrix}\right\}$$

Is the identity matrix, i.e.,

$$\begin{bmatrix} \langle u_1, u_1\rangle & \langle u_2, u_1\rangle \\ \langle u_1, u_2\rangle & \langle u_2, u_2\rangle \end{bmatrix} = \begin{bmatrix} 1 & 0 \\ 0 & 1 \end{bmatrix}$$

Let P be the transition matrix from the standar basis $\{e_1, e_2\}$ to the basis $\{u_1, u_2\}$, i.e,.

$$[u_1, u_2] = [e_1, e_2]P = [e_1, e_2]\begin{bmatrix} 2/3 & 1/3 \; 3 \\ 1/3 & -1/3 \end{bmatrix}$$

Let x' be the coordinate vector of the vector x relation basis B. (The coordinates vector of a x relative to the standards basis is itself x)

Then

$$x = [e_1, e_2]\, x = [u_1, u_2]\, x' = [e_1, e\ \ ]Px''$$

It follows that

$$x = Px'$$

Similarly, let y' be the coordinate vector of y relative to B. Then

$$y = Py'$$

Note that $x^T = x'^T P^T$. Thus, on the one hand, Theorem,

$$\langle x, y\rangle = x'^T I_n y' = x'^T y'$$

On the other hand,

$$\langle x, y\rangle = x^T A_y = x'^T P^T A P_{y'}$$

EXERCISES

Direction: Read the problems carefully and answer the questions that follow.

1. Show that the function that takes $((x_1, x_2), (y_1, y_2)) \in R^2 \, x \, R^2$ to $|x_1 y_1| + |x_2 y_2|$ is not an inner product on R^2.
2. Show that the function that takes $((x_1, x_2, x_3), (y_1, y_2, y_3)) \in R^3 \, x \, R^3$ to not an inner product on R^3.
3. Suppose $F = R \, and \, V \neq 0$. Replace the positivity condition (which states $\langle v, v \geq 0 \rangle$ for all $v \in V)$ in the definition of an inner product with the condition that $\langle v, v > 0 \rangle$ for some $v \in V$. Show that this change in the definition does not change the set of functions from V x V to R that are inner products on V.[1]
4. Suppose V is a real inner product space.
    1. Show that $\langle u + v, u - v = \|u\|^2 - \|v\|^2 \rangle$ for every u; $v \in V$.
    2. Show that if $u, v \in V$ have the same norm, then $u + v$ is orthogonal to u-v.
    3. Use part (b) to show that the diagonals of a rhombus are perpendicular to each other.[4]
5. Suppose V is finite-dimensional and $T \in LV/$ is such that $\|Tv\| \leq \|v\|$ for every $v \in V$. Prove that $T - \sqrt{2I}$ is invertible.
6. Suppose $u, v \in V$. Prove that $\langle u, v = 0 \rangle$ if and only if $\|u\| \leq \|u + av\|$ for all $a \in F$.

Direction: Read the problems carefully and answer the questions that follow.

1. Suppose $u, v \in V$ and $||u|| \leq 1$ if and $||v|| \leq 1$. Prove that $\sqrt{1 - ||u||^2}\sqrt{1 - ||v||^2 \leq 1 - |\langle u, v\rangle|}$

2. Prove that

$$(x_1 + \cdots + x_n)^2 \leq n\,(x_1^2 + \cdots + X_n^2)$$

   For all positive integers n and all real numbers $X_1, \ldots X_n$

3. Suppose $u, v \in V$. Prove that $||au + bv|| = ||bu + av||$ for all $a, b \in R$ if and only if

$$||a|| = ||u||$$

4. Suppose $u, v \in V$ and $||u|| = ||v|| = 1$ and $\langle u, v = 1\rangle$. Prove that u= u

5. Suppose $u, v \in V$ are such that

$$||u|| = 3, \qquad ||u + v|| = 4, \qquad ||u - v|| = 6$$

   What number does $||v||$ equal ?

6. Prove or disprove: there is an inner product on R^2 such that the associated norm is given by $||x.y|| \max\{|x|, |y|\}$

   For all $(x, y) \in R^2$

7. Suppose $p > 0$. Prove that there is an inner product on R^2 such that the associated norm is given by $||(x, y)|| = (|x|^p + |y|^p)^{1/p}$ for all $(x, y) \in R^2$ if and only if $p = 2$.

ORTHOGONAL COMPLEMENTS

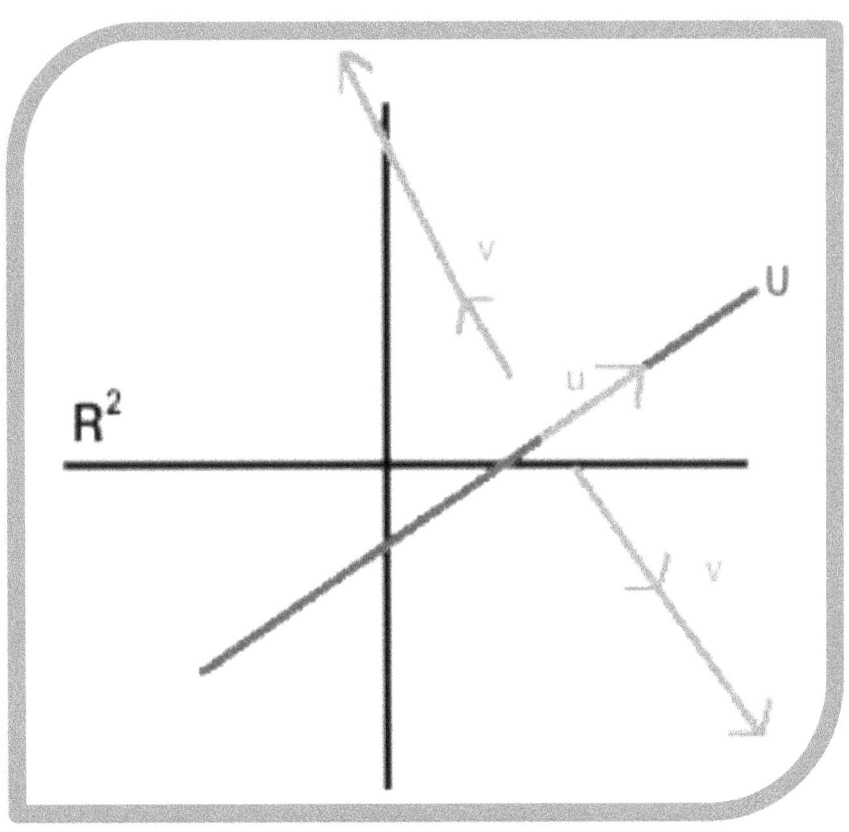

5.2 Orthogonal Complements

DISCUSSION

1. Orthogonality

Let V be an inner product space. Two vectors $u, v \in V$ are said to be **orthogonal** if [5]

$$\langle u, v\rangle = 0:$$

Example 1.1 For inner product space $C\,[-\pi, \pi]$, the functions *sin t* and *cos t* are orthogonal as [1]

$$\langle \sin t, \cos t\rangle = \int_{-\pi}^{\pi} \sin t \cos t \, dt$$

$$= \frac{1}{2} sin^2 t|_{-\pi}^{\pi} = 0 - 0 = 0$$

Example 1.2 Let $u = [a_1, a_2, \,::::\, a_n]^T \in R^n$. The set of all vectors of the Euclidean n- space R^n that are orthogonal to u is a subspace of R^n. It is the solution space of the single linear equation[1]

$$\langle u, v\rangle = a_1x_1 + a_2x_2 + \ldots + a_nx_n = 0$$

Let S be a nonempty subset of an inner product space V. We denote by $S^{\perp}$ the set of all vectors of V that are orthogonal to every vector of S, called the **orthogonal complement** of S in V. In notation,[1]

$S^{\perp} = \{v \in V | \langle u, v\rangle = 0 \, for \, all \, u \in S\}$

If S contains only one vector u, we write

$u^{\perp} = \{v \in V | \langle u, v\rangle = 0\}$

Let S be a nonempty subset of an inner product space. Then the orthogonal complement S⊥ is a subspace of V.

To show that S⊥ is a subspace. We need to show that S⊥ is closed under addition and scalar multiplication. Let u, v ∈ S⊥ and $c \in R$. Since $\langle u, w\rangle = 0$ and $\langle v, w\rangle = 0$ for all we S, then [1]

$$\langle u, +v, w\rangle = \langle u, w\rangle + \langle u, w\rangle = 0$$

$$\langle cu, w\rangle = c(u, w) = 0$$

For all $w \in S, u + v, cu$ ∈ S⊥ . Hence S⊥ is a substance of R^n.

Let S be a subset of an inner product space V. Then every vector of $\boldsymbol{S}\perp$ is orthogonal to every vector of Span (S), i.e., $\langle \boldsymbol{u}, \boldsymbol{v}\rangle = \mathbf{0}$, for all u ∈ Span (S), v ∈ $\boldsymbol{S}^{\perp}$.

For any $u \in Span\ (S)$, the vector u must be a linear combination of some vectors in S, say,

$$u = a_1u_1 + a_2u_2 + \ldots + a_ku_k$$

Then for any v ∈ S⊥

$$\langle u, v\rangle = a_1\langle u_1, v\rangle + a_2\langle u_2, v\rangle + \ldots + a_n(u_n, v) = 0$$

2. Orthogonal sets and bases

Let V be an inner product space. A subset S = $\{u_1, u_2, \ldots u_k\}$ of nonzero vectors of V is called an orthogonal set if every pair of vectors are orthogonal, i.e.,[1]

$$\langle u_i, u_j\rangle = 0, \quad 1 \le i \le k$$

An orthogonal set $S = \{u_1, u_2, \ldots u_k\}$ is called an orthogonal set if we further have

$$\|u_i\| = 0, \qquad 1 \le i < j \le k.$$

An orthogonal basis of V is a basis that is also an orthogonal set.

Theorem 7.1 (Pythagoras). Let $v_1, v_2, \dots\ v_k$ be mutually orthogonal vectors. Then $\|v_1 + v_2 + \cdots v_k\|^2 = \|v_1\|^2 + \|v_2\|^2 + \cdots + \|v_k\|^2$.

Proof. For simplicity, we assume k=2. If u and v are orthogonal, i.e. (u,v) = 0, then

$$\begin{aligned} \|u + v\|^2 &= (u + v, u + v) \\ &= \langle u, u\rangle + \langle v, v\rangle \\ &= \|u\|^2 + \|v\|^2 \end{aligned}$$

Example 2.1 The three vectors

$$v_1 = [1,2,1]^T, \quad v_2 = [2,1,-4]^T, \quad v_3 = [3,-2,1]^T$$

Are mutually orthogonal. Express the vector $v = [7,1,9]^T$ as a linear combination of v_1, v_2, v_3.

Set

$$x_1v_1 + x_2v_2 + x_3v_3 = v$$

There are two ways to find x_1, x_2, x_3

Method 1: Solving the linear system by performing row operations to tis augmented matrix [1]

$$[v_1, v_2, v_3 |\ v],$$

We obtain $x_1 = 3, x_2 = -1, x_3 = 2$. So, $v = 3v_1 - v_2 + 2v_3$

Method 2. Since $v_1 \perp v_j\ for\ i \neq j$, we have

$$\langle v, v_1\rangle = \langle x_1v_1 + x_2v_2 + x_3v_3, v_1\rangle = x_1\langle v_1, v_1\rangle$$

Where $i = 1\,;2\,;3$. Then

$$x_1 = \frac{\langle v, v_1 \rangle}{\langle v_1, v_1 \rangle}, \quad i = 1, 2, 3$$

We then have

$$x_1 = \frac{7+2+9}{1+4+1} = \frac{18}{6} = 3,$$

$$x_2 = \frac{14+1-36}{4+1+16} = \frac{-21}{21} = -1$$

$$x_3 = \frac{21-2+9}{9+4+1} = \frac{28}{14} = 2$$

3. GRAM-SCHMIDT PROCESS

Let W be a subspace of an inner product space V. Let $B = \{v_1, v_2, \dots v_k\}$ be a basis of W, not necessarily orthogonal. An orthogonal basis $B = \{w_1, w_2, \dots w_k\}$ may be constructed from B as follows: [1]

$w_1 = v_1$ $\qquad W_1 = Span\,\{w_1\},$

$w_2 = v_2 - Proj_{W_1}(v_2),$ $\qquad W_2 = Span\,\{w_1, w_2\},$

$w_3 = v_3 - Proj_{W_2}(v_3),$ $\qquad W_3 = Span\,\{w_1, w_2, w_3\},$

$w_{k-1} = v_{k-1} - Proj_{W_{k-1}}(v_{k-1}),$ $\qquad W_{k-1} = Span\,\{w_1, \dots, w_{k-1}\},$

$w_k = v_k - Proj_{W_{k-1}}(v_k)$

More precisely,

$w_1 = v_1,$

$$w_2 = v_2 - \frac{\langle w_1, v_2 \rangle}{\langle w_1, w_1 \rangle} w_1 - \frac{\langle w_2, v_k \rangle}{\langle w_2, w_2 \rangle} w_2 - \cdots - \frac{\langle w_{k-1}, v_k \rangle}{\langle w_{k-1}, w_{k-1} \rangle} w_{k-1}$$

The method of constructing the orthogonal vector $w_1, w_2, \dots, w_k$ is known as the **Gram-Schmidt process.**

Clearly, the vector $w_1, w_2, \dots, w_k$ are linear combinations of $v_1, v_2, \dots v_k$. Conversely, the Vectors $v_1, v_2, \dots v_k$ are also linear combinations of $w_1, w_2, \dots, w_k$.

$$v_1 = w_1,$$

$$v_2 = \frac{\langle w_1, v_2 \rangle}{\langle w_1, w_1 \rangle} w_1 + w_2,$$

$$v_3 = \frac{\langle w_1, v_3 \rangle}{\langle w_1, w_1 \rangle} w_1 + \frac{\langle w_2, v_3 \rangle}{\langle w_2, w_2 \rangle} w_2 + w_3,$$

$$v_k = \frac{\langle w_1, v_k \rangle}{\langle w_1, w_1 \rangle} w_1 + \frac{\langle w_2, v_k \rangle}{\langle w_2, w_2 \rangle} w_2 + \cdots + \frac{\langle w_{k-1}, v_k \rangle}{\langle w_{k-1}, w_{k-1} \rangle} w_{k-1} + w_k.$$

Hence

$$Span = \{v_1, v_2, \dots., v_k\} = Span\,\{w_1, w_2, \dots w_k\}$$

Since $B = \{v_1, v_2, \dots, v_k\}$ is a basis for W, so is the set $B' = \{w_1, w_2, \dots, w_k\}$.

Theorem 9.1. The basis $\{w_1, w_2, \dots, w_k\}$ constructed by the Gram-Schmidt process is an orthogonal basis of W. Moreover,

$$[v_1, v_2, v_{3,} \dots, v_k] = [w_1, w_2, w_3, \dots w_k]\, R,$$

Where R is the $k \; x \; k$ upper triangle matrix.

$$\begin{bmatrix} 1 & \frac{\langle w_1, v_2 \rangle}{\langle w_1, w_1 \rangle} & \frac{\langle w_1, v_3 \rangle}{\langle w_1, w_1 \rangle} & \cdots & \frac{\langle w_1, v_k \rangle}{\langle w_1, w_1 \rangle} \\ 0 & 1 & \frac{\langle w_2, v_3 \rangle}{\langle w_2, w_2 \rangle} & \cdots & \frac{\langle w_1, v_k \rangle}{\langle w_1, w_1 \rangle} \\ 0 & 0 & 1 & \cdots & \frac{\langle w_1, v_k \rangle}{\langle w_1, w_1 \rangle} \\ \vdots & \vdots & \vdots & \ddots & \vdots \\ 0 & 0 & 0 & \cdots & 1 \end{bmatrix}$$

Example 3.1. Let W be the subspace of R4 spanned by

$$v_1 = \begin{bmatrix}1\\1\\1\\1\end{bmatrix}, v_2 = \begin{bmatrix}1\\1\\1\\0\end{bmatrix}, \quad v_2 = \begin{bmatrix}1\\1\\0\\0\end{bmatrix}$$

Construct an orthogonal basis for W.

Set $w_1 = v_1$. Let $W_1 = Span\,\{w_1\}$. Ti find a vector w_2 in W that is orthogonal to W_1, set

$$w_2 = v_2 - Proj_{w_1}v_2 = v_2 - \frac{\langle w_1, w_2\rangle}{\langle w_1, w_1\rangle}w_1$$

$$= \begin{bmatrix}1\\1\\1\\0\end{bmatrix} - \frac{3}{4}\begin{bmatrix}1\\1\\1\\1\end{bmatrix} = \frac{1}{4}\begin{bmatrix}1\\1\\1\\-3\end{bmatrix}$$

Let $W_2 = Span\,\{W_1, W_2\}$. To find a vector w_3 in W that is orthogonal to W_2, set

$$w_3 = v_3 - Proj_{w_3}v_3$$

$$= v_3 - \frac{\langle w_1, v_3\rangle}{\langle w_1, w_1\rangle}w_1 - \frac{\langle w_2, v_3\rangle}{\langle w_2, w_2\rangle}w_2$$

$$= \begin{bmatrix}1\\1\\0\\0\end{bmatrix} - \frac{1}{2}\begin{bmatrix}1\\1\\1\\1\end{bmatrix} - \frac{\frac{2}{4}}{\frac{12}{16}}.\frac{1}{4}\begin{bmatrix}1\\1\\1\\-3\end{bmatrix}$$

$$= \begin{bmatrix}1/3\\1/3\\-2/3\\0\end{bmatrix}$$

Then the set $\{w_1, w_2, w_3\}$ is an orthogonal basis for W.

Theorem 9.2. Any m x n real matrix A can be written as

$$A = QR,$$

Called a QR – decomposition, where Q is an $m \, x \, n$ matrix whose columns are mutually orthogonal, and R is an $n \, x \, n$ upper triangle matrix whose diagonal entries are 1.

Example 3.2. Find a QR- decomposition of the matrix

$$A = \begin{bmatrix} 1 & 1 & 2 & 0 \\ 0 & 1 & 1 & 1 \\ 1 & 0 & 1 & 1 \end{bmatrix}$$

Let v_1, v_2, v_3, v_4 be the column vectors of A. Set

$$w_1 = v_1 = \begin{bmatrix} 1 \\ 0 \\ 1 \end{bmatrix}$$

Then $v_1 = w_1$. Set

$$w_2 = v_2 - \frac{\langle w_1, v_2 \rangle}{\langle w_1, w_1 \rangle} w_1$$

$$= \begin{bmatrix} 1 \\ 1 \\ 0 \end{bmatrix} - \frac{1}{2} \begin{bmatrix} 1 \\ 0 \\ 1 \end{bmatrix} = \begin{bmatrix} 1/2 \\ 1 \\ -1/2 \end{bmatrix}$$

Then $v_2 = (1/2) \, w_1 + w_2$. Set

$$w_3 = v_3 - \frac{\langle w_1, v_3 \rangle}{\langle w_1, w_1 \rangle} w_1 - \frac{\langle w_2, v_3 \rangle}{\langle w_2, w_2 \rangle} w_2$$

$$= \begin{bmatrix} 2 \\ 1 \\ 1 \end{bmatrix} - \frac{3}{2} \begin{bmatrix} 1 \\ 0 \\ 1 \end{bmatrix} - \begin{bmatrix} 1/2 \\ 1 \\ -1/2 \end{bmatrix} = 0$$

Then $v_3 = (3/2) w_1 + w_2 + w_3$. Set

$$w_4 = v_4 - \frac{\langle w_1, w_4\rangle}{\langle w_1, w_1\rangle} w_1 - \frac{\langle w_2, v_4\rangle}{\langle w_2, w_2\rangle} w_2$$

$$= \begin{bmatrix} 0 \\ 1 \\ 1 \end{bmatrix} - \frac{1}{2}\begin{bmatrix} 1 \\ 0 \\ 1 \end{bmatrix} - \frac{1}{3}\begin{bmatrix} 1/2 \\ 1 \\ -1/2 \end{bmatrix}$$

$$= \begin{bmatrix} -2/3 \\ 2/3 \\ 2/3 \end{bmatrix}$$

Then $v_4 = (1/2)w_1 + (1/3)w_2 + w_4$. Thus, matrixes Q and R for QR-decomposition of A are as follow [1]

$$Q = \begin{bmatrix} 1 & 1/2 & 0 & -2/3 \\ 0 & 1 & 0 & 2/3 \\ 1 & -1/2 & 0 & 2/3 \end{bmatrix}$$

$$R = \begin{bmatrix} 1 & 1/2 & 3/2 & 1/2 \\ 0 & 1 & 1 & 1/3 \\ 0 & 0 & 1 & 0 \\ 0 & 0 & 0 & 1 \end{bmatrix}$$

EXERCISES

Direction: Read the problems carefully and answer the questions that follow.

1. Let R^3 have the usual Euclidean inner product.

   a. Show that $\left\{\begin{bmatrix}1\\1\\1\end{bmatrix}, \begin{bmatrix}-1\\1\\0\end{bmatrix}, \begin{bmatrix}1\\2\\1\end{bmatrix}\right\}$ is a basis for R^3.

   b. Explain why this is not an orthogonal basis for R^3.

   c. Using the Gram-Schmidt process, transform this basis into an orthonormal basis for R^3.

2. Let R^4 have the usual Euclidean inner product.

   a. Show that $\left\{\begin{bmatrix}0\\2\\1\\0\end{bmatrix}, \begin{bmatrix}1\\-1\\0\\0\end{bmatrix}, \begin{bmatrix}1\\2\\0\\-1\end{bmatrix}, \begin{bmatrix}1\\0\\0\\1\end{bmatrix}\right\}$ is a basis for R^4

   b. Using the Gram-Schmidt process, transform this basis into an orthogonal basis for R^4.

3. Let R^3 have the inner product

$$\underset{v}{\rightarrow}.\underset{w}{\rightarrow} = v_1w_1 + 2v_2w_2 + 3v_3w_3$$

   1. Verify . is in fact an inner product of R^3.

   2. Use the Gram-Schmidt process to transform the basis

   $\left\{\begin{bmatrix}0\\2\\1\\0\end{bmatrix}, \begin{bmatrix}1\\-1\\0\\0\end{bmatrix}, \begin{bmatrix}1\\2\\0\\-1\end{bmatrix}, \begin{bmatrix}1\\0\\0\\1\end{bmatrix}\right\}$ into an orthogonal basis for R^3

4. Suppose $v_1, \ldots . v_m \in V$. Prove that

$$\{v_1, \ldots v_m\}^{\perp} = (\text{span}(v_1, \ldots v_m)^{\perp}$$

5. Suppose U is a finite-dimensional subspace of V. Prove that U⊥ = {0} if and only if U = V.

6. Suppose U is a subspace of V with basis $u_1, \ldots u_m$ and if and $u_1, \ldots u_{m,} W_n$ is a basis of V. Prove that if the Gram-Schmidt Procedure is applied to the basis of V above, producing a list $e_1, \ldots e_m, f_1, \ldots f_n$ then $e_1, \ldots e_m$ is an orthogonal basis of U and $f_1, \ldots f_n,$ then $e_1, \ldots e_m$ is an orthogonal basis of U and $f_1, \ldots f_n$ is an orthonormal basis of U⊥.

7. Suppose U is the subspace of R^4 defined by

$$U = span\left((1,2,3,-4),(-5,4,3,2)\right)$$

Find an orthogonal basis of U and an orthonormal basis of U⊥.

ASSESSMENT

Direction: Read the problems carefully and answer the questions that follow.

1. Perform the Gram-Schmidt process on each of these bases of R^3.

   a. $\langle \begin{pmatrix} 2 \\ 2 \\ 2 \end{pmatrix}, \begin{pmatrix} 1 \\ 0 \\ -1 \end{pmatrix}, \begin{pmatrix} 0 \\ 3 \\ 1 \end{pmatrix} \rangle$

   b. $\langle \begin{pmatrix} 1 \\ -1 \\ 0 \end{pmatrix}, \begin{pmatrix} 0 \\ 1 \\ 0 \end{pmatrix}, \begin{pmatrix} 2 \\ 3 \\ 1 \end{pmatrix} \rangle$

2. Perform the Gram-Schmidt process on each of these bases for R^2.

   a. $\langle (1), (1) \rangle$

   b. $\langle (-1), (-3) \rangle$

   c. $\langle (-1), (0) \rangle$

3. Find an orthogonal basis for this subspace of R^3: the plane $x - y + z = 0$.

4. Find an orthogonal basis for this substance of R^4.

$$\left\{ \begin{pmatrix} x \\ y \\ z \\ w \end{pmatrix} \middle| x - y - z + w = 0 \text{ and } x + z = 0 \right\}$$

5. Show that any linearly independent subset of R_n can be orthogonalized without changing its span.

6. What happens if we try apply the Gram-Schmidt process to a finite set that is not a basis?

7. What happens if we apply the Gram-Schmidt process to a basis that is already orthogonal?

8. Find a nonzero vector in R^3 that is orthogonal to both of these.

$$\begin{pmatrix} 1 \\ 5 \\ -1 \end{pmatrix} \begin{pmatrix} 2 \\ 2 \\ 0 \end{pmatrix}$$

Lesson 5.1 (Exercise)

2. The function $f\left((x_1, x_2, x_3), (y_1, y_2, y_3)\right) = x_1y_1 + x_3y_3$ is not an inner product of R^3 because for this function is not true that $\langle x, x\rangle = 0 \leftrightarrow x = (0,0,0)$ which is the property of definiteness.

4. a. Using the properties of the inner product, a function which has properties of positivity, definiteness, additivity in the first slot, homogeneity in the first slot, conjugate symmetry, we have that

$$\langle u + v, u - v\rangle = \|u\|^2 - \|v\|^2$$

b. If norms are equal, we have that $\langle u + v,\ u - v\rangle = 0$, which gives the fact that those vectors are orthogonal.

c. Because we prove that vector $u + v$ is orthogonal to $u - v$, because they have equal norms, we know that those vectors are perpendicular. Those vectors represent the diagonals of a rhombus.

5. We have that $\|Tv\| = |\lambda|\|v\| \leq \|v\|$ only if $|\lambda| \leq 1$, which is equivalent to the $\lambda \in [1,1]$. Because $\sqrt{2} \notin [1,1]$, we have that $\left|T - \sqrt{2I}\right| \neq 0$, so the matrix $T - \sqrt{2I}$ is invertible.

Lesson 5.2 (EXERCISES)

2. $\begin{bmatrix} 1 & -1 & 1 \\ 1 & 1 & 2 \\ 1 & 0 & 1 \end{bmatrix} \xrightarrow{rref} \begin{bmatrix} 1 & 0 & 0 \\ 0 & 1 & 1 \\ 0 & 0 & 1 \end{bmatrix}$

Since the rref ofthi matrix is the identity matrix, we conclude that the original set of vectors is a basis for R^3:

4. We have that for $v_i \in \{v_1, v_2, \dots v_m\}$ and for $w_j \in \{v_1, v_2, \dots v_m\}^{\perp}$ $\langle v_i, w_j\rangle = 0$, for $1 \leq i, j \leq m$.

That gives that $\{v_1, v_2, \dots, v_m\}^{\perp} = \left(span\left((v_1, v_2, \dots, v_m)\right)\right)^{\perp}$.

6. Because it is known that $V = U \oplus U^{\perp}$, and any $v \in V$ can be written as $v = \sum_{i=1}^{m} \alpha_i u_i + \sum_{j=1}^{n} \beta_j w_j$, we know that $\sum_{i=1}^{m} a_i u_i \in U$, so we get that $\sum_{j=1}^{n} \beta_j w_j \in U^{\perp}$.

   Now $\{e_1, \dots, e_m\}$ is an orthonormal basis of U and $\{f_{1,} \dots, fm\}$ is an orthonormal basis of $U^{\perp}$

CHAPTER 6

LINEAR TRANSFORMATION AND MATRICES

OBJECTIVES

After working through this section, you will be able to:

1. define linear transformations, one-to-one, onto and kernel;
2. verify if a given function is a linear transformation;
3. find the standard matrix representing the linear transformation;
4. verify if a linear transformation is one-to-one;
5. find ker L and dim ker L of a linear transformation;
6. verify if a linear transformation is onto;
7. find a basis for the range L;
8. find the matrix of a linear transformation.

As we have noted earlier, much of calculus deals with the study of properties of functions. Indeed, properties of functions are of great importance in every branch of mathematics, and linear algebra is no exception. In Chapter 1, wealready encountered functions mapping one vector space into another vector space; these are matrix transformations mapping R^n into R^m. Another example was isomorphisms between vector spaces, which we studied in Chapter 4. If we drop some of the conditions that need to be satisfied by a function on a vectorspace to be an isomorphism, we get another very useful type of function called a **linear transformation**.[1]

Linear transformations play an important role in many areas of mathematics, the physical and social sciences, and economics. A notation word: In Chapter 1 we denote a matrix transformation by *J*, and in Chapter 4 we denote an isomorphism by L. In this chapter, a function mapping one vector space into another vector space is denoted by L.[1]

6.1 Linear Transformation

DISCUSSION

1. Definition and Examples

Let V and W be vectors spaces. Afunstion $L: V\ WE$ is called a linear transformation of V into W if

a. L (u+v) = L (u) + L(v) for every u and v in V.

b. L (cu) =cL (u) in V, and ecu real number.

If V=W, the linear transformation $L: V \rightarrow W$ is also a linear operator j

Most of the vector spaces considered subsequently, but not all, are finite-dimensional.[1]

In the Definition above, observe that in (a) the + in **u + v** refers to the addition operation in V, whereas the + in *L (**u**) + L(**v**)* refers to the addition operation in W. Similarly, in (b) the scalar product *c**u*** is in V. while the scalar product *cL(**u**)* is in W.[1]

In the previous section, we have already pointed out that an isomorphism is alinear transformation that is one-to-one and onto. Linear transformations occur veryfrequently, and we now look at some examples.[2]

Example 1. Let A be an *m x n* matrix. We defined a matrix transformation as afunction

L: $R^n \rightarrow R^m$ defined *by L(u) = Au*. We now show that every matrix transformation is linear by verifying that properties (a) and (b) hold.

If *u* and *v* are vectors in R^n, then

L(u + v) = A(u + v) = Au + Av = L(u) + L (v).

Moreover, if *c* is a scalar, then

L(cu) = A(cu) = c(Au) = cL(u).

Hence, every matrix transformation is a linear transformation. •

For convenience, we now summarize the matrix transformations that have alreadybeen presented in the previous section.

Reflection with respect to the x-axis: L: $R^2 \rightarrow R^2$ is defined by

$$L\left(\begin{bmatrix} u_1 \\ u_2 \end{bmatrix}\right) = \begin{bmatrix} u_1 \\ -u_2 \end{bmatrix}$$

Projection into the xy-plane: $L: R^3 \rightarrow R^3$ is defined by

$$L\left(\begin{bmatrix} u_1 \\ u_2 \\ u_3 \end{bmatrix}\right) = \begin{bmatrix} u_1 \\ u_2 \end{bmatrix}$$

Dilation: $L: R^3 \rightarrow R^3$ is defined by $L\,(u) = ru\; for\; r > 1.$

Contraction: $L: R^3 \rightarrow R^3$ is defined by $L(u) = ru\; for\; 0 < r < 1.$

Rotation counterclockwise through an angle $\emptyset$: $L: R^2 \rightarrow R^2$

$$L\,(u) = \begin{bmatrix} \cos\emptyset & -\sin\emptyset \\ \sin\emptyset & \cos\emptyset \end{bmatrix} u$$

Example 2. Let $L: R^3 \rightarrow R^3$

$$L\left(\begin{bmatrix} u_1 \\ u_2 \\ u_3 \end{bmatrix}\right) = \begin{bmatrix} u_1 + 1 \\ 2u_2 \\ u_3 \end{bmatrix}$$

To determine whether L is a linear transformation, let

$$u = \begin{bmatrix} u_1 \\ u_2 \\ u_3 \end{bmatrix} \quad and \quad v = \begin{bmatrix} v_1 \\ v_2 \\ v_3 \end{bmatrix}$$

Then

$$L\,(u+v) = L\left(\begin{bmatrix}u_1\\u_2\\u_3\end{bmatrix}+\begin{bmatrix}v_1\\v_2\\v_3\end{bmatrix}\right) = L\left(\begin{bmatrix}u_1+\ v_1\\u_2+v_2\\u_3+v_3\end{bmatrix}\right) = \begin{bmatrix}(u_1+v_1)+1\\2(u_2+v_2)\\u_3+v_3\end{bmatrix}$$

On the other hand,

$$L(u+v) = \begin{bmatrix}u_1+1\\2u_2\\u_3\end{bmatrix}+\begin{bmatrix}v_1+1\\2v_2\\v\end{bmatrix} = \left[\begin{pmatrix}(u_1+v_1)+2\\2(u_2+u_2)\\u_3+v_3\end{pmatrix}\right]$$

Letting $u_1 = 1, u_2 = 3, u_3 = -3, v_1 = 2, v_2 = 4, v_3 = 1$, we see that $L = (u+v)\ \neq L(u) + L\,(v)$. Hence, we conclude that the function L is not a linear transformation.

Example 3. Let $L: R_2 \to R_2$ be defined by

$$L([u_1 \quad u_2]) = [u_1^2 \;\; 2u_2]$$

Is L is a linear transformation?

Solution

Let $u = [u_1 \quad u_2]\ \mathit{and} \quad v = [v_1 \quad v_2]$

Then

$$L\,(u+v) = L\,([u_1 \quad u_2]) + [v_1 \quad v_2]$$

$$= L\,([u_1+v_1 \quad u_2+v_2])$$

$$= [(u_1+v_1)^2 \quad 2(u_2+v_2)]$$

Since there are some choices of u and v such that $L\,(u+v) \neq L(u) + L(v)$. We conclude that L is not a linear transformation.[3]

Example 4. Let $L: P_1 \to P_2$ be defined by

$$L\,[p(t)] = tp(t).$$

Show that L is a linear transformation.

Solution

Let p(t) and q(t) be vectors in P_1 and let c be a scalar. Then

$$L\,[p(t)+q(t)] = t[p(t)+q(t)]$$

$$= tp(t)+tq(t)$$

$$= L\,[p(t)] + L[q(t)]$$

And

$$L[cp(t)] = t[cp(t)]$$

$$= c[tp(t)]$$

$$= cL[p(t)]$$

Hence L is a linear transformation.

Example 5. Let W be the vector space of all real-valued functions and let V be the subspace of all differentiable functions. Let $L: V \rightarrow W$ be defined by [3]

$$L(f) = f',$$

Where f' is the derivative of f. Using the properties of differentiation, we can show that L is a linear transformation.

Theorem 6.1

Let $L : V \rightarrow W$ be a linear transformation. Then

a. $L\ (0w) = 0w.$

$b. L\ (u - v) = L\ (u) - L(v), for\ u, v\ in\ V.$

Proof

a. We have

$0_v = 0_v + 0_v$

So

$L(0_v) = L\ (O_v + O_v)$

$L(O_v) = L\ (O_v) + L(O_v)$

Adding - $L\ (O_v)$ to both sides, we obtain

b. $L(u - v) = L(u + (-1)v) = L(u) + L\big((-1)v\big)$

$= L(u) + (-1)L\ (v) = L(u) - L(v)$

Example 6. Let V be an n-dimensional vector space and let $S = \{v_1, v_2, \dots, v_n\}$ an (3) in section 4.8 gives the relationship between the coordinate vector of v with respect to S and the coordinate vector of v with respect to T as

$$[v]_s = P_S \leftarrow T[v]_T.$$

where $P_{S \leftarrow T}$ is the transition matrix from T to S. Let L: $R^n \rightarrow R^n$ defined by

$$L(v) = P_{S \leftarrow T} v$$

For $v\ in\ R^n$. Thus L is a matrix transformation, so L is a linear transformation.

THEOREM 6.2

Let L: V → W be a linear transformation an n-dimensional vector space V into a vector space W. Let S = {vl, v2,..., vn} be a basis for V. If v is any vector in V, then L(v) is completely determined by

{L(v_1), L(v_2),..., L(v_n)}.

Proof

Since v is in V. we can write $v = a_1v_1 + a_2v_2 + ... + a_nv_n$ where a_1, a_2, . . . a_n are uniquely determinedreal numbers. Then

$$L(v) = L(a_1v_1 + a_2v_2 + ... + a_nv_n)$$
$$= L(a_1v_1) + L(a_2v_2) + ... + L(a_nv_n)$$
$$= a_1L(v_1) + a_2L(v_2) + ... + a_nL(v_n).$$

Thus L (v) has been completely determined by the vectors $L(v_1)$, $L(v_2)$, . . . , $L(v_n)$

Theorem 6.2 can also be stated in the following useful form: Let L: V → W and L' : V → W be lineartransformations of the n-dimensional vector space V into a vector space W. Let S = {vl, v2,..., vn} be a basis for V. If L' (v_i) = L(v_i) for i = 1, 2, ···, n, then L'(v) = L(v) for every v in V; that is, if Land L'agree on a basis for V, then L and L' are *identical linear transformations.*

Example 7. Let $L: R_4 \to R_2$ be a linear transformation and let $S =$ $= \{v_1, v_2, v_3, v_4\}$ be a basis for R_4, where $v_1 = [1 \quad 0 \quad 1 \quad 0]$, $v_2 =$ $[0 \quad 1 \quad -1 \quad 2]$, $v_3 = [0 \quad 2 \quad 2 \quad 1]$, $v_4 = [1 \quad 0 \quad 0 \quad 1]$

$L(v_1) = [1 \quad 2]$, $L(v_2) = [0 \quad 3]$,

$L(v_3) = [0 \quad 0]$, $L(v_4) = [2 \quad 0]$,

Let

$v = [3 \quad 5 \quad -5 \quad 0]$, Find L (v)

Solution

We first write v as a linear combination of the vectors in S, obtaining (verify)

$v = [3 \quad -5 \quad -5 \quad 0] = 2v_1 + v_2 - 3v_3 + v_4$

It then follows by Theorem 6.2 that

L(v) = $2v_1 + v_2 - 3v_3 + V_4$

$= 2L(v_1) + L(v_2) - 3L(v_3) + L(v_4) = [4 \quad 7].$

We already know if A is an m x n matrix, then the function L: $R^n \to R^m$ defined by $L(x) = Ax$ for x in $R^n$is a linear transformation. In the next example, we show that if $L: R^n \to R^m$ is a linear transformation, then L must be of this form. That is, L must be a matrix transformation.

Theorem 6.3

Let $L: R^n \to R^m$ be a linear transformation and consider the natural basis $\{e_1, e_2, \dots, e_3\}$ for R^n. Let A be the m x n matrix whose ith columns is $L\ (e_i)$. The matrix A has the following property:

If $x = \begin{bmatrix} x_1 \\ x_2 \\ \vdots \\ x_n \end{bmatrix}$ is any vector in R^n then $L\ (x) = Ax$ (equation 1)

Moreover, $A$is the only matrix satisfying Equation (1). It is called the **standard matrix representing L.**

Proof

Writing x as a linear combination of the natural basis for R^n, we have

$$x = x_1e_1 + x_2e_2 + \cdots + x_ne_n$$

So, by **Theorem 6.2**,

$$L(x) = L(x_1e_1 + x_2e_2 + \cdots + x_ne_n$$

$$= x_1L(e_1) + x_2L(e_2) + \cdots + x_nL(e_n) \qquad (equation\ 2)$$

Since A is the m x n matrix whose ${}_{\mathrm{i}}$th column is $L\left(e_j\right)$, we can write Equation (2) in matrix form as $L\,(x) = Ax$

Example 8. Let $L: R^3 \rightarrow R^2$ be a linear transformation defined by

$$L\left(\begin{bmatrix} x_1 \\ x_2 \\ x_3 \end{bmatrix}\right) = [x_1 + 2x_1^2 + 3x_2 + 2x_3]$$

Find the standard matrix representing L.

Solution.

Let $\{e_1, e_2, e_3\}$ be the natural basis for R^3. We now compute $L\left(e_j\right)$ for $j = 1,2,3$ as follows:

$$L\,(e_1) = L\begin{pmatrix} 1 \\ 0 \\ 0 \end{pmatrix} = \begin{bmatrix} 1 \\ 0 \end{bmatrix},$$

$$L\,(e_2) = L\begin{pmatrix} 0 \\ 1 \\ 0 \end{pmatrix} = \begin{bmatrix} 2 \\ 3 \end{bmatrix},$$

$$L(e_3) = L\begin{pmatrix} 0 \\ 0 \\ 1 \end{pmatrix} = \begin{bmatrix} 0 \\ -2 \end{bmatrix},$$

Hence, $A = [L(e_1)\ \ L(e_2)\ \ L(e_3)] = \begin{bmatrix} 1 & 2 & 0 \\ 0 & 3 & -2 \end{bmatrix}$

EXERCISES 6.1

Direction: Read the problems carefully and answer the questions that follow.

1. Determine whether the given function below is linear transformation.

   a) $L: R_3 \rightarrow R_3$ defined by
   $L([u_1 \quad u_2 \quad u_3]) = [u_1 \quad u_2^2 + u_3^2 \quad u_3^2]$

2. Is the given functions a linear transformation?

   a) $L: \quad \rightarrow \quad$ defined by
   $L([u_1 \quad u_2 \quad u_3]) = [0 \quad u_3 \quad u_2]$

3. Find the standard matrix representing the given linear transformation.

   a) $L: R^2 \rightarrow R^2$ defined by $L\left(\begin{bmatrix} u_1 \\ u_2 \end{bmatrix}\right) = \begin{bmatrix} -u_2 \\ u_1 \end{bmatrix}$

4. Find the standard matrix representing the given linear transformation.

For numbers 5-6. Let $A = \begin{bmatrix} 0 & -1 & 2 \\ -2 & 1 & 3 \\ 1 & 2 & -3 \end{bmatrix}$ be the standard matrix representing the linear transformation $L: R^3 \rightarrow R^3$.

5. Find $L\left(\begin{bmatrix} 2 \\ -3 \\ 1 \end{bmatrix}\right)$

6. Find $L\left(\begin{bmatrix} u_1 \\ u_2 \\ u_3 \end{bmatrix}\right)$

Direction: Read the problems carefully and answer the questions that follow.

Show your solution, if possible.

1. Which of the following functions are linear transformation?

   a) $L: R_2 \to R_3$ defined by
   $L([u_1 \quad u_2]) = [u_1 + 1 \quad u_2 \quad u_1 + u_2]$

   b) $L: R_3 \to R_3$ defined by
   $L([u_1 \quad u_2 \quad u_3]) = [1 \quad u_3 \quad u_2]$

   c) $L: R_2 \to R_3$ defined by
   $L([u_1 \quad u_2]) = [u_1 + 1 \quad u_2 \quad u_1 - u_2]$

2. Verify whether the given function is a linear transformation. [Here, p'(t) denotes the derivative of p(t) with respect to t.]

   a) $L: P_2 \to P_3$ by defined by $L(p(t)) = t^3 p'(0) + t^2 p(0)$.

3. Which of the following given functions is a linear transformation?

   a) $L: M_{nn} \to M_{nn}$ defined by $L(A) = A^T$
   b) $L: M_{nn} \to M_{nm}$ defined by $L(A) = A^{-1}$

4. Find the standard matrix representing each given linear transformation.

   a) $L: R^3 \to R^3$defined by $L\left(\begin{bmatrix} u_1 \\ u_2 \\ u_3 \end{bmatrix}\right) = \begin{bmatrix} u_1 \\ 0 \\ 0 \end{bmatrix}$

   b) $L: R^2 \to R^3$ defined by $L\left(\begin{bmatrix} u_1 \\ u_2 \end{bmatrix}\right) = \begin{bmatrix} u_1 - 3u_2 \\ 2u_1 - u_2 \\ 2u_2 \end{bmatrix}$

5. Verify whether the given function is a linear transformation. [Here, p'(t) denotes the derivative of p(t) with respect to t.]

   a) $L: P_1 \rightarrow P_2$ defined by $L(p(t)) = tp(t) + p(0)$.

6. Verify whether the given function below is a linear transformation. [Here, p'(t) denotes the derivative of p(t) with respect to t.]

   a) $L: P_1 \rightarrow P_2$defined by $L(p(t)) = tp(t) + 1$

6.2 Kernel and Range of a Linear Transformation

DISCUSSION

Definition 1

A linear transformation $L: V \rightarrow W$ is called **one-to-one** if it is a one-to-one function; that is, if $v_1 \neq v_2$ implies that $L(v_1) \neq L(v_2)$. An equivalent statement isthat L is one to-one if $L(v_1) = L(v_2)$ implies that $v_1 = v_2$.

EXAMPLE 1. Let L: $R^2 \rightarrow R^2$ defined by

$$L\left(\begin{bmatrix} u_1 \\ u_2 \end{bmatrix}\right) = \begin{bmatrix} u_1 + u_2 \\ u_1 - u_2 \end{bmatrix}$$

To determine whether L is one-to-one, we let

$v_1 = \begin{bmatrix} u_1 \\ u_2 \end{bmatrix}$ and $v_2 \begin{bmatrix} v_1 \\ v_2 \end{bmatrix}$

Then, if $L(v_1) = L(v_1)$, we have

$$u_1 + u_2 = v_1 + v_2$$

$$u_1 - u_2 = v_1 - v_2$$

Adding these equations, we obtain $2u_1 = 2v_1, or\ u_1 = v_1,$ which implies that $u_2 = v_2$.Hence, $v_1 = v_2$, and L is one-to-one.

Example: Let $L: R^3 \rightarrow R^2$ be the linear transformation defined by

$$L\left(\begin{bmatrix} u_1 \\ u_2 \\ u_3 \end{bmatrix}\right) = \begin{bmatrix} u_1 \\ u_2 \end{bmatrix}$$

Since $L\left(\begin{bmatrix}1\\3\\3\end{bmatrix}\right) = L\left(\begin{bmatrix}1\\3\\-2\end{bmatrix}\right)$, yet $\begin{bmatrix}1\\3\\3\end{bmatrix} \neq \begin{bmatrix}1\\3\\-2\end{bmatrix}$, we conclude that L is not one-to-one.

Definition 2

Let $L: V \rightarrow W$ be a linear transformation of a vector space V into a vector space
W. The **kernel** of L, *ker L*, is the subset of V consisting of all elements v of Vsuch that $L(v)$ = 0w.

We observe that *Theorem 6.1* assures us that *ker L* is never an empty set.because if $L: V \rightarrow W$ is a linear transformation, then 0v is in *ker L*.

Example. Let $L: R^3 \rightarrow R^2$ be as defined in Ex. 2. The vector $\begin{bmatrix}0\\0\\2\end{bmatrix}$

Is in *ker* L, since $L\left(\begin{vmatrix}0\\0\\2\end{vmatrix}\right) = \begin{vmatrix} \\ 0\end{vmatrix}$. However, the vector $\begin{bmatrix}2\\-3\\4\end{bmatrix}$ is not in ker L, since $L\left(\begin{vmatrix}2\\-3\\4\end{vmatrix}\right) = \begin{vmatrix} \\ -3\end{vmatrix}$. To find ker L, we must determine all v in R^3 so that $L(v) = 0_{R^2}$. That is, we seek $v = \begin{bmatrix}v_1\\v_2\\v_3\end{bmatrix}$ so that $L(v) =$

$$\left(\begin{bmatrix}v_1\\v_2\\v_3\end{bmatrix}\right) = \begin{bmatrix}0\\0\end{bmatrix} = 0_{R^2}.$$

However, $L(v) = [v_2]$. Thus, $[v_2] = \begin{bmatrix}0\\0\end{bmatrix}$, so $v_1 = 0, v_2 = 0$ *and* v_3 can be any real number. Hence, ker L consists of all vectors in R^3 of the form $\begin{bmatrix}0\\0\\0\end{bmatrix}$,

Where a is any real number. It is clear that ker L consists of the z-axis (x, y, z) three-dimensional space R^3.

THEOREM 6.4

Let $L: V \rightarrow W$ be a linear transformation of a vector space V into a vector space W. Then

(a) ker L is a subspace of V.
(b) L is one-to-one if and only if ker L = {Ov}.

EXAMPLE 4. Let L: P2 → R be the linear transformation defined by

$$L(at^2 + bt + c) = \int_0^1 (at^2 + bt + c)\, dt$$

(a) Find ker L.
(b) Find dim ker L.
(c) Is L one-to-one?

Solution

a) To find ker L, we seek an element $V = at^2 + bt + c$ in P_2 such that $L(v) = L\,(at^2 + bt + c) = 0_R = 0$, now

$$L(v) = \frac{at^3}{3} + \frac{bt^2}{2} + ct|_0^1 = \frac{a}{3} + \frac{b}{2} + c$$

Thus $c = -\frac{a}{3} - \frac{b}{2}$. Then ker L consists of all polynomials in P_2 of the form $at^2 + bt + \left(-\frac{a}{3} - \frac{b}{2}\right)$, for a and b any real numbers.

b) To find the dimension of ker L, we obtain a basis for ker L. Any vector ker L can be written as

$$at^2 + bt + \left(-\frac{a}{3} - \frac{b}{2}\right) = a\left(t^2 - \frac{1}{3}\right) + b\left(t - \frac{1}{2}\right)$$

Thus, the elements $\left(t^2 - \frac{1}{3}\right)$ and $\left(t - \frac{1}{2}\right)$ in P_2 span ker L. Now, these elements are also linearly independent, since they are not constant multiples of each other. Thus $\left\{t^2 - \frac{1}{3}, t - \frac{1}{2}\right\}$ is a basis ofr ker L, and dim ker L=2.

Definition 3

If $L: V \rightarrow W$ is a linear transformation of a vector space V into a vector space
W. then the range of L or image of V under L, denoted by range L, consists of allthose vectors in W that are images under L of vectors in V. Thus, w is in range Lif there exists some vector v in V such that $L(v) = w$. The linear transformation L is called ***onto*** if range $L = W$.

THEOREM 6.5

If $L: V \rightarrow W$ is a linear transformation of a vector space V into a vector spaceW, then range L is a subspace of W.

Proof

Let W1 and W2 be in range L. Then W1 = L(v1) and W2 = L(V2) for some V1 andV2 in V. Now

$$w_1 + w_2 = L(v_1) + L(v_2) = L(v_1 + v_2)$$

which implies that $w_1 + w_2$ is in range L. Also, if W is in range L, then $W = L(v)$
for some v in V. Then $cw = cL(v) = L(cv)$ where c is a scalar. so that cw is inrange L. Hence range L is a subspace of W.

c) Since dim ker L=2, L is not one-to-one.

Example 5. Consider Example 2 of this section again. Is this projection L onto?

Solution

We choose any vector $w = \begin{bmatrix} c \\ d \end{bmatrix}$ $in\ R^2$ and seek vector $v = \begin{bmatrix} v_1 \\ v_2 \end{bmatrix}$ $in\ V$ such that $L(v) = w$. Now $L(v) = \begin{bmatrix} v_1 \\ v_3 \end{bmatrix}$, so if $v_1 = c\ and\ v_2 = d, then\ L(v) = w$. Therefore, L is onto and dim range is $L = 2$.

Example 6. Consider Example 4 of this section, is L onto?

Solution

Given a vector W in R, $w = r$, a real number, can we find a vector $v = at^2 + bt + c$ in P_2 so that $L(v) = w = r$?

Now

$$L(v) = \int_0^1 (at^2 + bt + c)\, dt = \frac{a}{3} + \frac{b}{2} + c$$

We can let $a = b = 0$ and $c = r$. Hence L is onto. Moreover, dim range L=1.

Example 7. Let $L: R^3 \to R^3$ defined by

$$L = \left(\begin{bmatrix} u_1 \\ u_2 \\ u_3 \end{bmatrix}\right) = \begin{bmatrix} 1 & 0 & 1 \\ 1 & 1 & 2 \\ 2 & 1 & 3 \end{bmatrix} \begin{bmatrix} u_1 \\ u_2 \\ u_3 \end{bmatrix}$$

a) Is L onto?
b) Find a basis for range L.
c) Find ker L.
d) Is L one-to-one

Solution

a) Given any $w = \begin{bmatrix} a \\ b \\ c \end{bmatrix}$ in R^3, where a,b and c are any real numbers, can we find $v = \begin{bmatrix} v_1 \\ v_2 \\ v_3 \end{bmatrix}$ so that $L\ (v) = w?$ We seek a solution to the linear system.

$$\begin{pmatrix} 1 & 0 & 1 \\ 1 & 1 & 2 \\ 2 & 1 & 3 \end{pmatrix} \begin{bmatrix} u_1 \\ u_2 \\ u_3 \end{bmatrix} = \begin{bmatrix} a \\ b \\ c \end{bmatrix},$$

And we find the reduced row echelon form of the augmented matrix to be (verify)[3]

$$\begin{bmatrix} 1 & 0 & 1, & a \\ 0 & 1 & 1, & b - c \\ 0 & 0 & 0 & c - b - a \end{bmatrix}$$

Thus, a solution exists only for $c - b - a = 0$, so L is not onto.

b) To find a basis for range L, we note that

$$L = \left(\begin{bmatrix} v_1 \\ v_2 \\ v_3 \end{bmatrix}\right) = \begin{bmatrix} 1 & 0 & 1 \\ 1 & 1 & 2 \\ 2 & 1 & 3 \end{bmatrix} \begin{bmatrix} u_1 \\ u_2 \\ u_3 \end{bmatrix} = \begin{bmatrix} v_1 + v_3 \\ v_1 + v_2 + 2v_3 \\ 2v_1 + v_2 + 3v_3 \end{bmatrix}$$

$$= v_1 \begin{bmatrix} 1 \\ 1 \\ 2 \end{bmatrix} + v_2 \begin{bmatrix} 0 \\ 1 \\ 1 \end{bmatrix} + v_3 \begin{bmatrix} 1 \\ 2 \\ 3 \end{bmatrix}$$

This means that $\left\{\begin{bmatrix} 1 \\ 1 \\ 2 \end{bmatrix}, \begin{bmatrix} 0 \\ 1 \\ 1 \end{bmatrix}, \begin{bmatrix} 1 \\ 2 \\ 3 \end{bmatrix}\right\}$ spans range L. That is, range L is the subspace of R^3 spanned by the columns of the matrix defining L.

The first two vectors in this set are linearly independent since they are not constant multiples of each other. The third vector is the sum of the first two. Therefore, the first two vectors form a basis for range L, and dim range $L = 2.$ [3]

c) To find ker L, we wish to find all v in R^3 so that $L(v) = 0_R3$. Solving the resulting homogeneous system, we find (verify) that $v_1 = -v_3$ and $v_2 = -v_3$. Thus ker L consists of all vectors of the form.

$$\begin{bmatrix} -a \\ -a \\ a \end{bmatrix} = a \begin{bmatrix} -1 \\ -1 \\ 1 \end{bmatrix},$$

Where a is any real number. Moreover, dim ker $L = 1$

d) Since ker $L \neq \{0_R3$, it follows from Theorem 6.4(b) that L is not one-to-one.

EXERCISES 6.2

Direction: Read the problems carefully and answer the questions that follow.

1. Let $L: R^2 \to R^2$ be the linear transformation defined by $L\left(\begin{bmatrix} u_1 \\ u_2 \end{bmatrix}\right) = \begin{bmatrix} u_1 \\ 0 \end{bmatrix}$.

   a) Is $\begin{bmatrix} 0 \\ 2 \end{bmatrix}$ in ker *L*?
   b) Is $\begin{bmatrix} 3 \\ 0 \end{bmatrix}$ in range *L*?
   c) Is $\begin{bmatrix} 2 \\ 3 \end{bmatrix}$ in ker L?
   d) Is $\begin{bmatrix} 3 \\ 2 \end{bmatrix}$ in range *L*?
   e) Find *ker L*?
   f) Find range *L*?

2. Let $L: R^2 \to R^2$ be the linear transformation defined by $L\left(\begin{bmatrix} u_1 \\ u_2 \end{bmatrix}\right) = \begin{bmatrix} 1 & 2 \\ 2 & 4 \end{bmatrix}\begin{bmatrix} u_1 \\ u_2 \end{bmatrix}$.

   a) Is $\begin{bmatrix} 1 \\ 2 \end{bmatrix}$ in ker L?
   b) Is $\begin{bmatrix} 2 \\ -1 \end{bmatrix}$ is ker L?
   c) Is $\begin{bmatrix} 3 \\ 6 \end{bmatrix}$ in range L?
   d) Is $\begin{bmatrix} 2 \\ 3 \end{bmatrix}$ in range L?
   e) Find ker L.
   f) Find a set of vectors spanning range L.

3. Let $L: P_2 \to P_1$ be the linear transformation defined by $L(at^2 + bt + c) = (a + b)t + (b - c)$.

   a) Find a basis for ker L.
   b) Find a basis for range L.

Direction: Read the problems carefully and answer the questions that follow. Show your solutions, if possible.

1. Let $L: R_4 \rightarrow R_2$ be the linear transformation defined by

$$L\,([u_1 \quad u_2 \quad u_3 \quad u_4]) = [u_1 \quad u_2 \quad u_3 \quad u_4]$$

a) Is $[2 \quad 3 \quad -2 \quad 3]$ in ker L?
b) Is $[4 \quad -2 \quad -4 \quad 2]$ in ker L?
c) Is $[1 \quad 2]$ is range L?
d) Is $[0 \quad 0]$ in range L?
e) Find is L.
f) Find a set of vectors spanning range L.

2. Let $L: R_2 \rightarrow R_3$ be the lineart and transformation defined by
$$L\,([u_1 \quad u_2]) = [u_1 \quad u_2 \quad u_3 \quad u_4]$$
a) Find ker L.
b) Is L one-to-one?
c) Is L onto?

3. Let $L: R_4 \rightarrow R_3$ be the linear transformation define by

$$L\,([u_1 \quad u_2 \quad u_3 \quad u_4]) = [u_1 + u_2 \quad u_3 + u_4 \quad u_1 + u_3]$$

a.) Find the basis for ker L.
b.) What is dim ker L?
c.) Find a basis for range L.
d.) What is dim range L?

4. Find $L: R^5 \rightarrow R^4$ be the linear transformation defined by

$$L\left(\begin{bmatrix} u_1 \\ u_2 \\ u_3 \\ u_4 \\ u_5 \end{bmatrix}\right) = \begin{bmatrix} 1 & 0 & -1 & 3 & -1 \\ 1 & 0 & 0 & 2 & -1 \\ 2 & 0 & -1 & 5 & -1 \\ 0 & 0 & -1 & 1 & 0 \end{bmatrix} \begin{bmatrix} u_1 \\ u_2 \\ u_3 \\ u_4 \\ u_5 \end{bmatrix}$$

a.) Find the basis for and the dimension of ker L.

b.) What the basis for and the dimension of range L.

6.3 Matrix of a Linear Transformation

DISCUSSION

In the previous section, we saw that if A is an $m \times n$ matrix, we can define a linear transformation L: $R^n \to R^m$ by $L(x) = Ax$ for x in R^n. We shall now develop the following notion: If L: V → W is a linear transformation of an n-dimensional vector space V into an m-dimensional vector space W, and if we choose orderedbases for V and W, then we can associate a unique $m \times n$ matrix A with L that will enable us to find $L(x)$ for x in V by merely performing matrix multiplication.[4]

THEOREM 6.5

Let L: V→W be a linear transformation of an n-dimensional vector space V intoan m-dimensional vector space W ($n \neq 0$, $m \neq 0$) and let $S = \{v_1, v_2, \ldots, v_n\}$ and $T = \{w_1, w_2, \ldots, w_n\}$be ordered bases for V and W. respectively.

Then the $m \times n$ matrix A whose jth column is the coordinate vector $[L(v)_j]_T$ of $L(v)_j$
with respect to T has the following property:

$$[L(v)_j] = A[x]_S \text{ for every x in V.}$$

Moreover, A is the only matrix with this property.

Example 1.

(Calculus Required) Let L : P2 → P1 be defined by $L(p(t)) = p'(t)$, andconsider the ordered bases $S = \{t^2, t, 1\}$ and $T = \{t, 1\}$ for P2 and P1, respectively.

(a) Find the matrix A associated with L.

(b) If $p(t) = 5t^2 - 3t + 2$, compute L(p(t)) directly and then by using A.

Solution

a. We have

$$L(t^2) = 2t = 2t + 0(1), \qquad so\ [L(t^2)]_T = \begin{bmatrix} 2 \\ 0 \end{bmatrix}$$

$$L(t^2) = 1 = 0(t) + 1(1), \qquad so\ [L(t)]_T = \begin{bmatrix} 0 \\ 1 \end{bmatrix}$$

$$L(1) = 0 = 0(t) + 0(1), \qquad so\ [L(1)]_T = \begin{bmatrix} 0 \\ 10 \end{bmatrix}$$

In this case, the coordinates of L(t2), L(t), and L(1) with respect to the T-basisare obtained by observation since the T - basis is quite simple. Thus

$$A = \begin{bmatrix} 2 & 0 & 0 \\ 0 & 1 & 0 \end{bmatrix}$$

(b) Since $p(t) = 5t^2 - 3t + 2$, then $L(p(t)) = 10t - 3$. However, we can find $L(p(t))$ by using the matrix A as follows: Since

$$[p(t)]_s = \begin{bmatrix} 5 \\ -3 \\ 2 \end{bmatrix}$$

Then

$$[L(p(t))]_T = \begin{bmatrix} 2 & 0 & 0 \\ 0 & 1 & 0 \end{bmatrix} \begin{bmatrix} 5 \\ -3 \\ 2 \end{bmatrix} = \begin{bmatrix} 10 \\ -3 \end{bmatrix}$$

This means that $L(p(t)) = 10t - 3$

Example 2. Let $L: P_2 \to P_1$be defined as in Example 1 and consider the ordered bases $S = \{1, t, t^2\}$ and $T = \{t, 1\}$ for P_2 and P_1, respectively. We then find that the matrix A associated with $L\ is\ \begin{bmatrix} 0 & 0 & 2 \\ 0 & 1 & 0 \end{bmatrix}$ (verify). Notice that if we change the order of the vectors in S or T, the matrix may change.[5]

Example 3. Let $L: P_2 \to P_1$ be defined as in Example I, and consider the ordered bases $S = \{t^2, t, 1\}$ and $T = \{t, +1, t-1\}$ for P_2 and $P_{1,}$respectively.

(a) Find the matrix A associated with L.

(b) If $p(t) = 5t^2 - 3t + 2$, compute $L(p(t))$.

Solution

a. We have

$$L(t^2) = 2t$$

To find the coordinates of $L\ (t^2)$ with respect to the T- basis, we form $L(t^2) = 2t = \ a_1(t+1) + \ a_2 - 1,$ Which leads to the linear system

$$a_1 + a_2 = 2$$

$$a_1 - \ a_2 = 0,$$

Whose solution is $a_1 = 1, a_2 = 1$. Hence

$$[L(t^2)]_T = \begin{bmatrix} 1 \\ 1 \end{bmatrix}$$

Similarly,

$L(t) = 1 = \frac{1}{2}\ (t+1) - \frac{1}{2}(t-1),$ so $[L(t)]_T = \begin{bmatrix} \frac{1}{2} \\ \frac{-1}{2} \end{bmatrix}$

$L(t) = 0 = 0(t+1) + 0(t-1),$ so $[L(1)]_T = \begin{bmatrix} 0 \\ 0 \end{bmatrix}$

Hence $A = \begin{vmatrix} 1 & \frac{1}{2} & 0 \\ 1 & -\frac{1}{2} & 0 \end{vmatrix}$

b. We have

$$[L(p(t))]_T = \begin{bmatrix} 1 & \dfrac{1}{2} & 0 \\ 1 & -\dfrac{1}{2} & 0 \end{bmatrix} \begin{bmatrix} 5 \\ -3 \\ 2 \end{bmatrix} = \begin{bmatrix} \dfrac{7}{2} \\ \dfrac{13}{2} \end{bmatrix},$$

So $L(p(t)) = \frac{7}{2}(t+1) + \frac{13}{2}(t-1) = 10t - 3$, which agrees with the results found in

Example 1.

Notice that the matrices obtained in Examples I, 2, and 3 are different, even though L is the same in all three examples. In the next section, we will discuss therelationship between any two of these three matrices.[5]

The matrix A is called **the representation of L with respect to the ordered bases S and T**. We also say that **A represents L with respect to Sand T**. HavingA enables us to replace L by A and x by [x]s to get A [x]s = [L(x)]T'. Thus, applying L to X in V to obtain L (x) in W can be found by multiplying the matrix A by the matrix [x]s. That is, we can work with matrices rather than with linear transformations. Physicists and others who deal at great length with linear transformations perform most of their computations with the matrix representationsof the linear transformations. Of course, it is easier to work on a computer with matrices than with our abstract definition of a linear transformation. The relationship between linear transformations and matrices is a much stronger one than mere computational convenience. In the next section, we show that the set of all linear transformations from an n-dimensional vector space V to an m-dimensionalvector space W is a vector space that is isomorphic to the vector space Mn of allm x n matrices.[5,6]

We might also mention that if L: $R^n \rightarrow R^m$ is a linear transformation, then we often use the natural bases for R^n and R^m, which simplifies obtaining a representation of L.

Example 4. Let $L: R^3 \rightarrow R^2$ be defined by

$$L\left(\begin{bmatrix} x_1 \\ x_2 \\ x_3 \end{bmatrix}\right) = \begin{bmatrix} 1 & 1 & 1 \\ 1 & 2 & 3 \end{bmatrix}\begin{bmatrix} x_1 \\ x_2 \\ x_3 \end{bmatrix}$$

Let

$$e_1 = \begin{bmatrix}1\\0\\0\end{bmatrix},\quad e_2 = \begin{bmatrix}0\\1\\0\end{bmatrix},\quad e_3 = \begin{bmatrix}0\\0\\1\end{bmatrix},$$

$$e_1 = \begin{bmatrix}1\\0\end{bmatrix},\qquad e = \begin{bmatrix}0\\1\end{bmatrix}$$

Then $S = \{e_1, e_2, e_3\}$ and $T = \{\bar{e_1}, \bar{e_2}\}$ are the natural bases for R^3 and R^2, respectively.

Now

$$L(e_1) = \begin{vmatrix}1 & 1 & 1\\1 & 2 & 3\end{vmatrix}\begin{vmatrix}1\\0\\0\end{vmatrix} = \begin{vmatrix}1\\1\end{vmatrix} = 1e_1 + 1e_2,\qquad \text{so}[L(e_1)]_T = \begin{bmatrix}1\\1\end{bmatrix},$$

$$L(e_2) = \begin{vmatrix}1 & 1 & 1\\1 & 2 & 3\end{vmatrix}\begin{vmatrix}0\\1\\0\end{vmatrix} = \begin{vmatrix}1\\2\end{vmatrix} = 1e_1 + 2e_2,\quad \text{so } [L(e_2)]_T = \begin{bmatrix}1\\2\end{bmatrix},$$

$$L\,(e_3) = \begin{vmatrix}1 & 1 & 1\\1 & 2 & 3\end{vmatrix}\begin{vmatrix}0\\0\\1\end{vmatrix} = \begin{vmatrix}1\\3\end{vmatrix} = 1e_1 + 2e_2\quad \text{so } [L(e_3)]_T = \begin{bmatrix}1\\3\end{bmatrix},$$

In this case, the coordinate vectors of $L(e_1), L(e_2)$ *and* $L(e_3)$ with respect to theT-basis are readily computed because T is the natural basis for R^2. Then the representation of L with respect to S and T is

$$A = \begin{bmatrix}1 & 1 & 1\\1 & 2 & 3\end{bmatrix}$$

A is the same matrix as the one involved in the definition of L because the natural bases are being used for R^3 and R^2.

Example 5. . Let L: $R^3 \to R^2$ be defined as in Example 4, and consider the ordered bases

$$S = \left\{\begin{bmatrix}1\\1\\0\end{bmatrix}, \begin{bmatrix}0\\1\\1\end{bmatrix}, \begin{bmatrix}0\\0\\1\end{bmatrix}\right\} \textbf{\textit{and } } \boldsymbol{T} = \left\{\begin{bmatrix}1\\2\end{bmatrix}, \begin{bmatrix}1\\3\end{bmatrix}\right\}$$

For R^3 and R^2, respectively. Then

$$L\left(\begin{bmatrix}1\\1\\0\end{bmatrix}\right)=\begin{bmatrix}1&1&1\\1&2&3\end{bmatrix}\begin{bmatrix}1\\1\\0\end{bmatrix}=\begin{vmatrix}2\\3\end{vmatrix}$$

Similarly,

$$L\left(\begin{bmatrix}0\\1\\1\end{bmatrix}\right)=\begin{vmatrix}2\\5\end{vmatrix}\qquad and\ L\left(\begin{bmatrix}0\\0\\1\end{bmatrix}\right)=\begin{bmatrix}1\\3\end{bmatrix}$$

To determine the coordinates of the images of the S-basis. We must solve the three linear systems

$$a_1\begin{bmatrix}1\\2\end{bmatrix}+a_3\begin{bmatrix}1\\3\end{bmatrix}=b$$

where $b=\begin{vmatrix}2\\3\end{vmatrix},\begin{vmatrix}2\\5\end{vmatrix}\ and\ \begin{vmatrix}1\\3\end{vmatrix}$. This can be done simultaneously as discussed in the previous section, by transforming the partitioned matrix

$$\begin{bmatrix}1&1&2&2&1\\2&3&3&5&3\end{bmatrix}$$

To reduce row echelon form, yielding (verify)

$$\begin{bmatrix}1&0&3&1&0\\0&1&-1&1&1\end{bmatrix}$$

The last three columns of this matrix are the desired coordinate vectors of the image of the S-basis with respect to the T -basis. That is, the last three columnsform the matrix A representing L with respect to Sand T.[5] Thus,

$$A=\begin{bmatrix}3&1&0\\-1&1&1\end{bmatrix}$$

This matrix, of course, differs from the one that defined L. Thus, although amatrix A may be involved in the definition of a linear transformation L, we cannot conclude that it is necessarily the representation of L that we seek.

From Example 5 we see that if L: Rn→ Rm is a linear transformation. then a computationally efficient way to obtain a matrix representation A of L with respect to the ordered bases $S = \{v1, v2, . . . , vn\}$ for R'' and $T = \{w1, w2, . . . , wn\}$

for R^m is to proceed as follows: Transform the partitioned matrix

$[w_1 \quad w_2 \quad ... \quad w_n : L(v_1) : L(v_2) : ... : L(v_n)]$ to reduced row echelon form. The matrix A consists of the last *n* columns of this last matrix.[5]

If $L : V \to V$ is a linear operator on an n-dimensional space V, then to obtaina representation of L, we fix ordered bases S and T for V, and obtain a matrix.

A represent L with respect to S and T. However, it is often convenient in this case to choose S = T. To avoid verbiage in this case; we refer to A as the ***representation of L with respect to S.*** . For example, if L: $R^n \to R^m$ is a linear operator, then the matrix representing L with respect to the natural basis for R^n has already been discussed in Theorem 6.3 in Section 6.1 where it was called ***the standard matrix representing L.***

EXERCISES 6.3

Direction: Read carefully and answer the questions that allow.

1. Let $L: R^2 \rightarrow R^2$ be defined by $L\begin{pmatrix}u_1\\u_2\end{pmatrix} = \begin{bmatrix}u_1 + 2u_2\\2u_1 - u_2\end{bmatrix}$. Let S be the natural basis for R^2 $and\ let\ T = \left\{\begin{bmatrix}-1\\2\end{bmatrix}, \begin{bmatrix}2\\0\end{bmatrix}\right\}$

   Find the representation of L with respect to:

   $a.)\ S;$ $b.)\ S\ and\ T$

   $c.)\ T\ and\ S;$ $d.)\ T;$

e) Find $L\left(\begin{bmatrix}u_1\\u_2\end{bmatrix}\right)$ by using the definition of L and also by using the matrices found a. thorough (d)

2. Let $L: R_4 \rightarrow R_3$ be defined by $L([u_1 \quad u_2 \quad u_3 \quad u_4]) = [u_1 \quad u_2 + u_3 \quad u_3 + u_4]$. Let S and T be the natural bases for $R_1\ and\ R_{3,}$ respectively. Let

$$S' = \{[1 \quad 0 \quad 0 \quad 1], [0 \quad 0 \quad 0 \quad 1], [1 \quad 1 \quad 0 \quad 0], [0 \quad 1 \quad 1 \quad 0]\}$$

And

$$T' = \{[1 \quad 1 \quad 0], [0 \quad 1 \quad 0], [1 \quad 0 \quad 1]\}$$

a Find the representation of L with respect to S and T.

b. Find the representation of L with respect to S' and T'

c. Find $L = ([2 \quad 1 \quad -1 \quad 3])$ by using the matrices obtained in parts (a) and (b)and compare this answer with that obtained from the definition for L.

3. Let $L: R^3 \rightarrow R^3$ be defined by

$$L\left(\begin{bmatrix}1\\0\\0\end{bmatrix}\right) = \begin{bmatrix}1\\1\\0\end{bmatrix}, \quad L\left(\begin{bmatrix}0\\1\\0\end{bmatrix}\right) = \begin{bmatrix}2\\0\\1\end{bmatrix}, \quad L\left(\begin{bmatrix}0\\0\\1\end{bmatrix}\right) = \begin{bmatrix}1\\0\\1\end{bmatrix}$$

a.) Find the representation of L with respect to the natural basis S and R^3.

b.) Find $L\left(\begin{bmatrix}1\\2\\3\end{bmatrix}\right)$ by using the definition of L and also by using the matrix obtained in part (a).

4. Let L: $R3 \rightarrow R3$ be defined in Exercise 5. Let T = $\{(L(e_1), (L(e_2)$, $(L(e_3)\}$ be nan ordered basis for R3, and let S be the natural basis for R_3.

a.) Find the representation of L with respect to S and T.

b.) Find $L\left(\begin{bmatrix}1\\2\\3\end{bmatrix}\right)$ by using the matrix obtained in part (a).

ASSESSMENT

Direction: Read the problems carefully and answer the questions that follow.Show your solutions, if possible.

1. Let $L: R^4 \rightarrow R^4$ be defined by $L\left(\begin{bmatrix} u_1 \\ u_2 \\ u_3 \\ u_4 \end{bmatrix}\right) = \begin{bmatrix} 1 & 0 & 1 & 1 \\ 0 & 1 & 2 & 1 \\ -1 & -2 & 1 & 0 \end{bmatrix}\begin{bmatrix} u_1 \\ u_2 \\ u_3 \\ u_4 \end{bmatrix}$

Let S and T be the natural bases for R^4 and R^3, respectively, and consider the orderedbases

$$S' = \left\{\begin{bmatrix} 1 \\ 1 \\ 0 \\ 0 \end{bmatrix}, \begin{bmatrix} 0 \\ 1 \\ 0 \\ 0 \end{bmatrix}, \begin{bmatrix} 0 \\ 0 \\ 1 \\ 1 \end{bmatrix}, \begin{bmatrix} 0 \\ 1 \\ 1 \\ 0 \end{bmatrix}\right\} \text{ and } T' = \left\{\begin{bmatrix} 1 \\ 0 \\ 1 \end{bmatrix}, \begin{bmatrix} 0 \\ 1 \\ 1 \end{bmatrix}, \begin{bmatrix} 0 \\ 0 \\ 1 \end{bmatrix}\right\}$$

for R^4 and R^3, respectively. Find the representation of L with respect to:

a.) S and T   b.) S' and T'

2. Let L: $R^2 \rightarrow R^2$ be the linear transformation rotating R^2 counterclockwise through anangle Ø. Find the representation of L With respect to the natural basis for R^2.

3. Let $L: R^3 \rightarrow R^3$ be the linear transformation represented by the matrix $\begin{bmatrix} 1 & 3 & 1 \\ 1 & 2 & 0 \\ 0 & 1 & 1 \end{bmatrix}$ with respect to the natural basis for R^3. Find:

   a) $L\left(\begin{bmatrix} 1 \\ 2 \\ 3 \end{bmatrix}\right)$ b) $L\left(\begin{bmatrix} 0 \\ 1 \\ 1 \end{bmatrix}\right)$

4. (Calculus Required) Let V be the vector space with basis $S = \{1, t, e', te'\}$ and let $L: V \rightarrow V$ be a linear operator defined by L(f) = f '= df/dt. Find the representationof L with respect to S.

KEY TO CORRECTION

Exercises 6.1

2. YES

4. $\begin{bmatrix} 0 & -1 \\ -1 & 0 \end{bmatrix}$

6. $\begin{bmatrix} -x_2 + 2x_3 \\ -2x_2 + x_2 + 3x_3 \\ x_1 + 2x_2 - 3x_3 \end{bmatrix}$

Exercise 6.2

2. a) NO
   b) YES
   c) YES
   d) NO
   e) All vectors of the form $\begin{bmatrix} -2a \\ a \end{bmatrix}$, where a is any real number
   f) A possible answer is $\left\{\begin{bmatrix} 1 \\ 2 \end{bmatrix}, \begin{bmatrix} 2 \\ 4 \end{bmatrix}\right\}$

4. a) $\{-t^2 + t + 1\}$
   b) $\{t, 1\}$

Exercise 6.3

2. a) $\begin{bmatrix} 1 & 0 & 0 & 0 \\ 0 & 1 & 1 & 1 \\ 0 & 0 & 1 & 1 \end{bmatrix}$ b) $\begin{bmatrix} 0 & -1 & 1 & -1 \\ 0 & 1 & 0 & 3 \\ 1 & 1 & 0 & 1 \end{bmatrix}$

   c) $\begin{bmatrix} 2 & 0 & 2 \end{bmatrix}$

4. a) $\begin{bmatrix} 1 & 0 & 0 \\ 0 & 1 & 0 \\ 0 & 0 & 1 \end{bmatrix}$ b) $\begin{bmatrix} 1 \\ 2 \\ 3 \end{bmatrix}$

Notes

Introduction

1. 1.Linear Algebra (2nd Edition) - PDF Free Download. https://epdf.pub/linear-algebra-2nd-edition.html
2. Jim Hefferon - Joshua. https://joshua.smcvt.edu/linearalgebra/book_ebook.pdf

Chapter 1

1. 1RWIRU6DOH 4 Equations; Matrices Systems of Linear. https://www.pearsonhighered.com/assets/samplechapter/0/3/2/1/0321947622.pdf
2. Linear Algebra - Term Paper. https://www.termpaperwarehouse.com/essay-on/Linear-Algebra/297696
3. 4_linear_equations.pdf - MATHEMATICS FOR DATA MANAGEMENT https://www.coursehero.com/file/123109945/4-linear-equationspdf/
4. Linear algebra 3e - Tài liệu text. https://text.123docz.net/document/4820019-linear-algebra-3e.htm
5. Introduction to Linear Algebra. http://www.voutsadakis.com/TEACH/LECTURES/LINEAR/Chapter1.pdf
6. 1 Linear Equations. https://math.berkeley.edu/~arash/54/notes/01_01.pdf
7. 8.1(1).pdf - Section 8.1 Systems of Linear Equations in https://www.coursehero.com/file/81749520/811pdf/
8. Systems of Linear Equations in Two Variables. https://www.regent.edu/app/uploads/2019/06/ML-Math-102-Systems-of-Equations.pdf
9. 1. Mathematical concepts. 2. Basics concepts in Quantum .... https://web2.aabu.edu.jo/tool/course_file/lec_notes/403741_Lect%201%20revision%20of%20QM%20new.pdf
10. If ra=sa+t, then a?. https://study-assistantph.com/math/question513711744
11. Transfer Matrix Analysis. https://core.ac.uk/download/pdf/53186907.pdf
12. Composition of two matrices | Physics Forums. https://www.physicsforums.com/threads/composition-of-two-matrices.779108/
13. Module in Linear Algebra and Matrix Theory. https://studylib.net/doc/6880997/module-in-linear-algebra-and-matrix-theory
14. Matrices 1 - University of Nebraska–Lincoln. http://www.math.unl.edu/~scohn1/EngRevf08/matrix

15. 1. Mathematical concepts. 2. Basics concepts in Quantum .... https://web2.aabu.edu.jo/tool/course_file/lec_notes/403741_Lect%201%20revision%20of%20QM%20new.pdf

Chapter 2

1. WEEK3.pdf - 1 To see how a linear system whose augmented https://www.coursehero.com/file/75081134/WEEK3pdf/
2. Dr Abdul Hanan Sheikh. http://ahsheikh.github.io/Courses/LinAlg/echelon_form_slides.pdf
3. Nsm. Nsm
4. Systems of Equations and Inequalities Copyright Cengage .... https://slidetodoc.com/systems-of-equations-and-inequalities-copyright-cengage-learning/
5. Some Linear Algebra Notes. https://people.math.wisc.edu/~meyer/math340/340notes.pdf
6. c Type III Add a multiple of one row column to another 52 .... https://www.coursehero.com/file/p19casks/c-Type-III-Add-a-multiple-of-one-row-column-to-another-52-Learning-Module-in/
7. WEEK3.pdf - 1 To see how a linear system whose augmented https://www.coursehero.com/file/75081134/WEEK3pdf/
8. LEARNING GAUSS-JORDAN ELIMINATION USING MS EXCEL.. https://123dok.com/document/q2m34j6y-learning-gauss-jordan-elimination-using-ms-excel.html
9. Math 211. https://math.rice.edu/~polking/slides/fall02/lecture19bw3.pdf
10. SOLVED:If A and B are nonsingular matrices, then (A B)^{T https://www.numerade.com/questions/if-a-and-b-are-nonsingular-matrices-then-a-bt-is-nonsingular-and-lefta-btright-1lefta-1righttleftb-2/
11. Module in Linear Algebra and Matrix Theory. https://studylib.net/doc/6880997/module-in-linear-algebra-and-matrix-theory

Chapter 3

1. Module in Linear Algebra and Matrix Theory. https://studylib.net/doc/6880997/module-in-linear-algebra-and-matrix-theory
2. Some Linear Algebra Notes. https://people.math.wisc.edu/~meyer/math340/340notes.pdf
3. Elementary Linear Algebra with Applications (9th Edition https://silo.pub/elementary-linear-algebra-with-applications-9th-edition.html
4. PPT - 資訊科學數學 14 : Determinants & Inverses PowerPoint https://www.slideserve.com/miracle/14-determinants-inverses

5. 340notes - University of Wisconsin–Madison. https://people.math.wisc.edu/~meyer/math340/Notes.html
6. Next we compute the determinant of the upper triangular https://www.coursehero.com/file/p7qtn9k5/Next-we-compute-the-determinant-of-the-upper-triangular-matrix-30-det-2-1-8-120/

Chapter 4

1. Linear Algebra - PDF Free Download - Donuts. https://epdf.pub/linear-algebra93294cf5ae9d294109f5ef4209910dd284693.html
2. Linear Algebra (2nd Edition) - PDF Free Download. https://epdf.pub/linear-algebra-2nd-edition.html
3. Linear Algebra for Competitive Exams. Linear Algebra for Competitive Exams
4. LINEAR ALGEBRA I. http://sncollegenattika.ac.in/admin/uploads/LINEAR%20ALGEBRA%20by%20DIVIYA%20K%20D%20(1).pdf
5. Algebra Linear - Hoffman - Excelente livro, porém possui https://www.docsity.com/pt/algebra-linear-hoffman/4896463/
6. Linear Algebra - UFPE. https://cin.ufpe.br/~jrsl/Books/Linear%20Algebra%20-%20Kenneth%20Hoffman%20&%20Ray%20Kunze%20.pdf
7. 1 Preliminaries. http://www2.aueb.gr/users/demos/vectors.pdf
8. Algebra Linear - Hoffman - Excelente livro, porém possui https://www.docsity.com/pt/algebra-linear-hoffman/4896463/
9. Vector Spaces. https://sam.nitk.ac.in/courses/MA904/vector%20spaces.pdf

Chapter 5

1. Inner Product Spaces - Axler. https://linear.axler.net/InnerProduct.pdf
2. Linear Algebra (2nd Edition) - PDF Free Download. https://epdf.pub/linear-algebra-2nd-edition.html
3. Vector Spaces - CSU. https://www.math.colostate.edu/~renzo/teaching/M369-F2017/Chapter5-annotated.pdf
4. (c) Use part (b) to show that the diagonals of a rhombus https://www.numerade.com/questions/suppose-v-is-a-real-inner-product-space-a-show-that-langle-uv-u-vrangleu2-v2-for-every-u-v-in-v-b-sh/
5. Two vectors are said to be orthogonal if.docx - Two https://www.coursehero.com/file/95196323/Two-vectors-are-said-to-be-orthogonal-ifdocx/

Chapter 6

1. mm11(1).docx - MODULE 5 LINEAR TRANSFORMATIONS AND https://www.coursehero.com/file/115109232/mm111docx/
2. Two vectors are said to be orthogonal if.docx - Two https://www.coursehero.com/file/95196323/Two-vectors-are-said-to-be-orthogonal-ifdocx/
3. Linear Transformation-Linear Algebra-Handout - Docsity. https://www.docsity.com/en/linear-transformation-linear-algebra-handout/171072/
4. Some Linear Algebra Notes. https://people.math.wisc.edu/~meyer/math340/340notes.pdf
5. Elementary Linear Algebra with Applications (9th Edition https://silo.pub/elementary-linear-algebra-with-applications-9th-edition.html
6. Operator is it function? | Physics Forums. https://www.physicsforums.com/threads/operator-is-it-function.40173/

References

Bernard Kolman and David R. Hill .Elementary Linear Algebra with Application ninth edition

Bernard Kolman, et. al. (2008) Elementary Linear Algebra with applications,9th Ed. Pearson Education, Inc,.

Bernard Kolman, et. al. (2008) Instructor's Solutions Manual Elementary LinearAlgebra with Applications, 9th Ed. Pearson Education, Inc,.

BUSINESS MATHEMATICS and STATISTICS by Dr. J.K. THUKRAL

David Cherney, et. al. (2013) Linear Algebra, 1st Ed. Davis California.

Dr. P C Tulsian & Bharat Jhunjhunwala. Quantitative Aptitude for CPT, 1/e. S Chand Publishing.

Hill https://www.mathplanet.com

Hoffman, Kenneth et. al. Linear Algebra, 2nd Ed. PRENTICE-HALL, INC. , Englewood Cliffs, New Jersey

http://joshua.smcvt.edu/linearalgebra
https://linear.axler.net/InnerProduct.pdf

http://www.math.sci.hokudai.ac.jp/~s.settepanella/teachingfile/Algebra/Algebra2/pa gine/Lalg2_prova_middle_sol.pdf

https://courses.lumenlearning.com/ivytech-collegealgebra/chapter/solving-a-system- of-linear-equations-using-matrices/

https://en.wikipedia.org/wiki/Linear_algebra

https://textbooks.math.gatech.edu/ila/row.reduction.html

https://www.math.colostate.edu/~clayton/teaching/m369s17/exams/exam2practices olutions.pdf

https://www.math.tamu.edu/~dallen/linear_algebra/chpt3.pdf

https://www.mathplanet.com

https://www.mathsisfun.com/algebra/-determinants.html

https://www.mathsisfun.com/algebra/-solving-linear-systems.html

https://www.mathsisfun.com/algebra/systems-linear-equations-matrices.html

https://www.open.ac.uk/courses/modules/m208

https://www.shelovesmath.com/algebra/advanced-algebra/solving-systems-using- reduced-row-echelon-form/

https://www.slader.com/textbook/9783319110790-linear-algebra-done-right- third edition

https://www.statisticshowto.com/matrices-and-matrix-algebra/reduced-row-echelon- form-2/

www.ingramcontent.com/pod-product-compliance
Ingram Content Group UK Ltd.
Pitfield, Milton Keynes, MK11 3LW, UK
UKHW021934200726
13853UKWH00011B/2071